Super-diversity increasingly cha
world. *Global Cities, Local Streets*
tion of the ways growing diversification plays out in the ecosystems of
shopping districts and everyday experiences of shopkeepers and shoppers. The
richly described cases and compelling theoretical insights give us a timely, new
understanding of contemporary urban transformations.

*Steven Vertovec, Max Planck Institute for the Study of Religious and Ethnic
Diversity*

Beyond the apparently banal frame of "shopping streets" lies a dense network
of expansive and parochial practices of exchange. *Global Cities, Local Streets*
offers a compelling comparison across six cities, advancing new insights into
the substance and methods of transnational research. This book is a fine-
grained contribution to the field of global urbanisation, and will be an
invaluable teaching resource.

*Suzanne Hall, Sociology & LSE Cities, London School of Economics and Political
Science*

Focusing on local shopping streets in global cities as diverse as Amsterdam to
Tokyo, this book examines the way the intimate and the transactional, the
neighbourly and the far-flung, the familiar and the strange, as well as identity,
belonging, moral ownership, and social mobility, come together in constitut-
ing the life-worlds of local shopping streets. Offering a richly textured,
curb-side view of the way everyday civility, conviviality, cosmopolitanism, and
conflict play out, we are reminded that neither gentrification nor ghettoization
are inexorable processes. Instead, kaleidoscopic diversity, shifting at every turn,
seems to be that which nourishes and sustains the streets of our times.

Brenda Yeoh, Geography, National University of Singapore

Global Cities, Local Streets

Everyday Diversity from New York to Shanghai

Global Cities, Local Streets: Everyday Diversity from New York to Shanghai, a cutting-edge text/ethnography, reports on the rapidly expanding field of global, urban studies through a unique pairing of six teams of urban researchers from around the world. The authors present shopping streets from each city—New York, Shanghai, Amsterdam, Berlin, Toronto, and Tokyo—how they have changed over the years, and how they illustrate globalization embedded in local communities. This is an ideal addition to courses in urbanization, consumption, and globalization.

The book's companion website, www.routledge.com/cw/zukin, has additional videos, images, and maps, alongside a forum where students and instructors can post their own shopping street experiences.

Sharon Zukin is professor of sociology at Brooklyn College and the Graduate Center of the City University of New York, and was a visiting professor in the Center for Urban Studies at the University of Amsterdam in 2010–11. She has written three books about New York City: *Loft Living, The Cultures of Cities*, and *Naked City: The Death and Life of Authentic Urban Places*, as well as *Point of Purchase: How Shopping Changed American Culture*. She won the C. Wright Mills Award from the Society for the Study of Social Problems for her book *Landscapes of Power: From Detroit to Disney World*.

Philip Kasinitz is presidential professor of sociology at the Graduate Center of the City University of New York. His books include *Caribbean New York: Black Immigrants and the Politics of Race, Metropolis: Center and Symbol of Our Time, Becoming New Yorkers: Ethnographies of The New Second Generation*, and *The Urban Ethnography Reader*. He is co-author of *Inheriting the City: The Children of Immigrants Come of Age*, which received the 2010 Distinguished Publication Award from the American Sociological Association.

Xiangming Chen is the dean and director of the Center for Urban and Global Studies and Paul E. Raether distinguished professor of global urban studies and sociology at Trinity College in Hartford, Connecticut. He is also a distinguished guest professor in the School of Social Development and Public Policy at Fudan University in Shanghai. His (co-)authored and co-edited books include *As Borders Bend: Transnational Spaces on the Pacific Rim, Shanghai Rising: State Power and Local Transformations in a Global Megacity*, and *Rethinking Global Urbanism: Comparative Insights from Secondary Cities*.

Titles of Related Interest from Routledge

Point of Purchase: How Shopping Changed American Culture
Sharon Zukin

Foodies: Democracy and Distinction in the Gourmet Foodscape, Second Edition
Josée Johnston and Shyon Baumann

Race, Space, and Exclusion: Segregation and Beyond in Metropolitan America
Edited by Robert Adelman and Christopher Mele

Informal Urban Street Markets: International Perspectives
Edited by Clifton Evers and Kirsten Seale

Global Cities, Local Streets

Everyday Diversity from New York to Shanghai

Sharon Zukin, Philip Kasinitz, and Xiangming Chen

Routledge
Taylor & Francis Group

NEW YORK AND LONDON

First published 2016
by Routledge
711 Third Avenue, New York, NY 10017

and by Routledge
2 Park Square, Milton Park, Abingdon, Oxon, OX14 4RN

Routledge is an imprint of the Taylor & Francis Group, an informa business

© 2016 Taylor & Francis

Library of Congress Cataloging in Publication Data
Global cities, local streets: everyday diversity from New York to Shanghai / edited by Sharon Zukin, Philip Kasinitz, and Xiangming Chen. — 1 Edition.
pages cm
Includes bibliographical references and index.
1. Gentrification. 2. Culture and globalization. I. Zukin, Sharon, editor.
HT170.G5594 2015
303.48'2—dc23
2015003267

ISBN: 978-1-138-02392-5 (hbk)
ISBN: 978-1-138-02393-2 (pbk)
ISBN: 978-1-315-77619-4 (ebk)

Typeset in Minion Pro
by FiSH Books Ltd, Enfield

Contents

List of Figures

List of Tables

CHAPTER 1

Spaces of Everyday Diversity
The Patchwork Ecosystem of Local Shopping Streets

SHARON ZUKIN, PHILIP KASINITZ, AND XIANGMING CHEN

If you want to see the diversity that is driving the growth of cities today, take a walk on the shopping street of almost any neighborhood. These local streets are fast becoming a "global" urban habitat, where differences of language and culture are seen, heard, smelled, felt, and certainly tasted. Here is where globalization is embedded in local communities, where immigrants from different regions of the world work alongside the native-born, and the national dishes of foreign cuisines, from pizza to *pupusas*, become local attractions.

Whether we're walking, shopping, taking our clothes to the dry cleaner, or getting a bite to eat, these are the spaces where we experience everyday diversity.

Yet local shopping streets are the most taken-for-granted spaces on the planet. Surrounded by houses and dotted with small stores, they seem like useful but insignificant passageways between our homes and the wider world. But they are not only places for economic exchange. Local shopping streets express an equally important need for social sustainability and cultural exchange.

Where do they come from? How do they change? What does diversity on a local shopping street really mean?

Though they are less famous than the central agora of ancient Athens or the forum of ancient Rome, local shopping streets have equally historic roots (see Figure 1). This kind of street creates a miniature marketplace for nearby residents and forms a "natural" community center. It is often a hot spot of urban vitality. Yet today, in an age of accelerated mobility and global "flows,"

Figure 1 Ruins of Olive Oil Shop, Pompeii, Destroyed by Volcanic Eruption in Year 79
Source: Photo by Richard Rosen.

local shops risk losing their livelihood to both suburban shopping malls and online retail sales.

We who put this book together believe that local shopping streets should thrive. We like their human scale, the way they enable social interaction, no matter how fleeting or superficial, and their environmental friendliness because they encourage shoppers to walk or bike to do errands. We also like locally owned stores because the money they make stays in the city instead of flowing out to distant corporate headquarters. Moreover, there is a good chance that their owners feel an empathy with the street, their customers, and the neighborhood. We admire the communities that are revitalized and sustained by small store owners, who are often migrants from rural areas or overseas.

With these points in mind, we bring you the life stories of twelve local shopping streets in six global cities around the world, traveling from New York to Shanghai by way of Toronto, Amsterdam, Berlin, and Tokyo. We find remarkable similarities in the way they work and the risks they face, and also outstanding differences in the way they respond to two major challenges of our time, globalization and gentrification.

Looking at multiple sites around the world gives us a broad perspective to answer questions of concern to everyone who cares about cities and diversity. In what sense does a seemingly random assortment of local shops not only embody a city's past, but also its future?

A Social World

Let's begin with the positive things we find in most local shopping streets, including the twelve we write about. Grocery stores and takeout delis, dry cleaners, hair salons: clustered together, local shops make urban life possible by offering city dwellers a convenient place to get the goods and services they need to survive. But local shops also make city life sociable (see Figure 2).

Think about cafés, bars, barber shops, and nail salons: people spend time in these places, exchanging gossip and news, or maybe just saying "good morning" when they buy a cup of coffee, making a momentary connection to both the wider world and their home community. Both inside the store and outside in the street, local shops sustain social interaction.

Look at the shopkeepers who sweep the sidewalk and keep an eye on passersby. City laws usually require them to keep the sidewalk clean, but they do so much more.

The perceptive urban writer Jane Jacobs (1961) noted years ago how shopkeepers protect the social order of the street. They watch out for crimes, offer school children a safe haven inside their shops, and create an island of familiarity in a world of strangers. On the street where Jacobs lived in New York City, shopkeepers knew many neighbors' names, accepted packages for them if they were not at home when deliveries arrived, and kept an extra set of their apartment keys for emergencies. Taking on these unpaid responsibilities, business owners and their employees provided local residents with both safety and convenience.

Yet despite shopkeepers' involvement in their customers' daily lives, the local shopping street in Jacobs's city is not a traditional village "where everybody knows your name." Most of her shopkeepers did not live in the neighborhood, and while many shoppers did, the social life of the street did not exclude outsiders.

Figure 2   Elizabeth Street, Manhattan, Old Storefronts of Italian-American Community, in Year 2000
Source: Photo by Richard Rosen.

For Jacobs, the local shopping street is a distinctly urban space that is neither as intimate as the home nor as anonymous as the central business district. At its best, this kind of space provides for the needs of both neighbors and strangers.

In Jacobs's time, before the era of shopping malls, superstores, and online shopping began, city dwellers could satisfy most of their daily needs on their local shopping street. Each store—"the butcher, the baker, the candlestick maker," in the words of an old nursery rhyme—specialized in a different task. All over the city, local shopping streets replicated the same specialized functionality. Yet each neighborhood's special character, its "DNA," was encoded in the *ecosystem* of its local shopping street.

If an *ecosystem* is a complex network with many interrelated parts, all interacting with the surrounding environment, the ecosystem of a local shopping street brings together in one compact physical space the networks of social, economic, and cultural exchange created every day by store owners, their employees, shoppers, and local residents. These networks may be as far-flung as the global migrations that bring men and women to open *taquerias* and Chinese restaurants in cities in the Global North, and as local as customers from the next block who come to the small shop, greengrocer, or *bodega* which is still open at midnight to buy a container of orange juice.

Ideally, to satisfy everyday needs, you never have to leave your neighborhood.

Shopping Streets and Neighborhoods

In the 1940s, the writer E. B. White described New York as a city of "countless small neighborhoods," where each local shopping street offered residents the means to be "virtually self-sufficient." Every local shopping street offered a wide range of goods and services, some for every day and others for less frequent needs. Certain kinds of businesses have become obsolete since then. But when White recounts the types of local shops he knows, the names strike a rhythm that still sounds familiar. "No matter where you live in New York," he says,

> you will find within a block or two a grocery store, a barbershop, a newsstand and shoeshine shack, an ice-coal-and-wood cellar … a dry cleaner, a laundry, a delicatessen (beer and sandwiches delivered at any hour to your door), a flower shop, an undertaker … a movie house, a radio-repair shop, a stationer, a haberdasher [men's clothing store], a tailor, a drugstore, a garage, a tearoom, a saloon [bar], a hardware store, a liquor store, a shoe-repair shop (White 1949: 35–37).

Likewise in Shanghai, before the 1949 revolution, local shops form the same kind of neighborhood ecosystem, which embodies the DNA of the local community:

Among the most common stores in [the] neighborhoods were those that sold grain, coal, cotton fabrics and goods, groceries, hot water, condiments, snacks, fruit, wine, meat and vegetables, and other products. Other shops offered such services as tailoring, barbering, repair of household items, and currency exchange, and there also were laundries, tea houses, and public bathhouses. In short, in Shanghai's *lilong* neighborhoods [where small houses were built around alleyways], the merchandise and services most closely related to daily life could be purchased within a block of one's home (Lu 1995).

Like New York and Shanghai, all global cities develop the same reiterative ecosystem, an endlessly repeated patchwork of retail stores and services. Most of them were, and still are, individually or family-owned, "mom-and-pop" businesses.

Looking at this history, we can easily feel nostalgic for an imagined lost intimacy of local life. But we can also see some of the objective factors that still make local shopping streets a significant part of the city's social, cultural, and environmental ecosystem: walkability, interdependence, a diversity of goods and services, and the opportunity to make connections. These patterns are both disrupted and enhanced by new technologies and migrations.

New Technologies and Migrations

Today, people enjoy a greater choice of products but require more geographical mobility to consume them. They also rely on more media of social communication to find out where to buy specific goods and services, compare prices, and read reviews. Without thinking too much about it, shoppers have become dependent on automobiles, electricity, and electronic devices, as well as on global markets and journeys. Not surprisingly, the interaction between technologies and migrations has reshaped local shopping streets.

Families now have refrigerators and sometimes even freezers in their home, so they don't have to go shopping every day for food. Many men and women own cars, so they can travel farther to search for goods with lower prices or better quality. Other transportation innovations have broadened the range of shopping options, starting with container shipping and air freight, which bring goods from faraway regions. But they often bring them to supermarkets, big box discount stores, and suburban shopping malls, where low prices, greater selection, and more convenient opening hours may lure customers away from local shops.

Though these global developments have led to the decline of many local shopping streets, or to what in the UK is called "the crisis of the High Street" (Portas 2011), others have gained both vitality and variety from transnational migrations. Fresh fruits and vegetables from Asia, Africa, or the Caribbean are sold by greengrocers who themselves have recently migrated from those regions. Individual shops, and the street as a whole, host an unprecedented range of mixtures and fusions. On one local shopping street in Manhattan, the owner of the dry cleaners comes from Korea while the employee who works at the counter comes from Honduras and the tailor who sews at a table in the corner is Chinese. Chinese migrants run small shops in remote corners of Africa, as do North Africans in Paris, West Africans in Brooklyn, and Vietnamese in Prague.

The products sold in local shopping streets, and the shopkeepers who sell them, are often the first forms of globalization city dwellers consciously encounter.

But technology has shifted many kinds of consumption away from human interaction in brick-and-mortar stores. While online companies like Amazon and Apple have expanded from selling books and computers to music, groceries, and household appliances, shoppers have also got used to buying more individualized items like clothes and shoes from retail websites. In Brooklyn, young professionals use apps to order laundry service instead of taking their wash to the local laundromat.

On the other hand, good reviews on social media bring new customers to local shops and restaurants. Other apps help small store owners to manage inventory, process credit card payments, and create video advertisements, if they have the time and skills to use them.

All in all, despite the lure of the distant and the new, local shops are still important to urban life. They are built to a human scale and create redundant sources of supply, producing both a socially sustainable *habitus* and an environmentally sustainable *habitat*.

But to know how they work, we must understand their ecosystem. Where do local shopping streets come from? What do they do? How do they interact with larger changes in the city, namely, globalization and gentrification?

A Globalized Habitus

Sociologists and anthropologists use the term *habitus* to indicate a set of everyday practices and aesthetic tastes that are shared by social and cultural groups who socialize together—and socialize each other (Bourdieu 1984, 1990). We can think of local shopping streets as a *habitus* in two senses, as both a "conceptual" space, embodying, reproducing, and symbolizing the collective tastes of a social group, and a "lived" space, which is physical, functional, and experiential (adapted from Lefebvre 1991). The conceptual space is the one we visualize when we think of the shopping street. And the lived space is where we go shopping.

At its best, as both a conceptual space and a lived one, a local shopping street can be safe and inclusive. It can provide a safe space for encounters with the new and different. But at its worst, it can be dangerous and segregated by race, ethnicity, wealth, or gender. As Jacobs warned, when a local shopping street is the uncontested "turf" of some social groups, it risks becoming a space of exclusion for others.

Let's take food as an example. Food is a common currency of globalization, circulating cultural goods and practices from different regions of the world among "natives" and migrants from different areas.

Whether they sell Polish *pierogi* (dumplings), Salvadorian *pupusas* (meat or cheese pastries), or bagels baked according to a recipe imported from Montreal, local food shops create a small but significant space of multicultural sociability. Shoppers who are only strangers may lay down their suspicions when they are shopping for food; they interact peaceably or at least shop side by side, and accept each other with some degree of conviviality, civility, and maybe even empathy (Amin 2012; Anderson 2011; Hall 2012).

Of course, it doesn't always happen this way. Some people prefer to maintain more insular patterns of consumption. Even those who develop a more cosmopolitan palate will not necessarily apply this attitude to other people and other spheres of social life. But more often than not, food shopping provides a safe encounter with unfamiliar others. And the city is usually the better for it.

It's not just what is sold that brings together the global and the local. It is also the sellers, both shopkeepers and their employees. In many cities, the businesses on local shopping streets are mainly owned and run by migrants. If in earlier times migrant shopkeepers came from small towns in nearby provinces, now they often travel a greater distance, across national borders and oceans.

Small retail stores can provide an entry point into the economy for men and women who migrate with little capital or education. Family members may work in the shop, reducing the need to pay wages. Merchandise may be supplied, sometimes on credit, by co-ethnic networks of wholesalers and dealers. Financial costs to open a small shop, particularly in a working-class neighborhood, are generally low, and rents are even lower for merchants who sublet a small space in a store from another business owner who may come from the same village or country (Gold 2010; Min 2011). It's not unusual in New York, London, or Amsterdam to see a notary public from Ghana or Pakistan sharing a storefront with a travel agent, jeweler, and vendor of DVDs, all of them transnational migrants.

Sometimes these businesses start out by selling goods and services to co-ethnics. Demographic changes in the surrounding residential neighborhood may create a business opportunity for immigrant merchants to provide both products from "back home" and the things a growing migrant community needs to survive in the new land. Over time a concentration of such businesses may reshape the habitus of local shopping streets, making it noticeably more "global," or making it global in different ways.

With transnational migration and local settlement, clusters of ethnic businesses form Little Italys, Little Havanas, and Little Senegals. And while these clusters usually begin by serving migrant communities, they may be discovered by adventurous members of other groups as well, including food shoppers looking to satisfy new tastes.

From Ethnic Clusters to Super-diversity

In U.S. cities, especially in New York, ethnic clusters on local shopping streets are seen as an amenity, and are celebrated for adding vitality and diversity to the larger urban landscape (Hum 2014; Lin 2010; Taylor 2000). They may even be marketed as a tourist attraction (Conforti 1996; Rath 2007). But in some European cities, they are seen as a divisive "balkanization" and a threat to social cohesion (Hall 2015). Public officials who try to break up ethnic clusters in Amsterdam, for example, claim that they do so in the name of "diversity," which has the opposite meaning from "diversity" in New York because it brings in more members of the ethnic majority, native-born Dutch, instead of more members of ethnic minorities.

But ethnic clusters do not last forever. When an ethnic group or immigrant community moves away from the neighborhood, or the initial cohort of store owners ages, retires, and passes from the scene, their business enclave will shrink. In rare cases, such as New York's Little Italy, the cluster may endure as an ethnic theme park, retaining the "ethnic flavor" even though the ethnicity of the shopkeepers has changed (Kosta 2014). Often, the shopkeepers' children will prefer going to college to standing behind a counter. If they all get this opportunity, the ethnic cluster of their parents' shops will eventually disappear.

This is what happened to the enclave of Jewish-owned stores on New York's Lower East Side, where the first generation of street vendors and shopkeepers arrived from Russia and Eastern Europe in the 1880s. Their stores are gone, for the most part, today. It is worth noting, however, that businesses opened by the first cohort of these Jewish shopkeepers took the place of shops and bars that had been established by earlier German immigrants, in a process of ethnic succession.

As a result of these changes, longtime local residents may feel alienated from the neighborhood when the shops connected with their cultural or ethnic traditions close or move away. It's not only that they can no longer buy familiar goods from people with familiar faces. Disappearance of "their" shops changes the habitus of the local shopping street, and group members gradually lose their sense of moral ownership of the surrounding territory.

Stores owned by global migrants are not limited to ethnic enclaves, or shopping streets in immigrant neighborhoods. In many cities migrant shopkeepers serve native populations, sometimes using their family and community networks to develop an ethnic niche in a specific economic sector. In that case, native-born customers come to think of the local greengrocers or nail salons as "Korean" or "Turkish" businesses because of the owners' ethnicity.

Migrants and ethnic "outsiders" are often shopkeepers in low-income, work-ing-class, and racial-minority neighborhoods. While some may be well liked by their customers, the fact that they are not co-ethnics can lead to deep resentment and feelings of powerlessness in the surrounding community. "We don't even control the businesses in our own neighborhood!" is a persistent cry in many African-American communities, where "middleman minorities" have become targets of local frustration and even violence (Gold 2010; Kasinitz and Haynes 1996; Min 1996). Yet though riots make the headlines, overt conflicts between shopkeepers and local residents are rare, even in racially polarized neighborhoods (Lee 2002).

Today, in many cities around the world, a new pattern is developing in which immigrant shopkeepers are neither concentrated in ethnic clusters like a Little Italy or Little Senegal, nor serve native populations as middleman minorities. Instead, the new globalized habitus is that of a "super-diverse" local shopping street, in which both shopkeepers and customers come from a wide variety of different, though still predominantly immigrant, back-grounds.

This new local habitus reflects the "super-diversity" of the global city, where no ethnic group holds a clear majority and geographical communities are made up of a wide variety of men and women from different national origins, of different social backgrounds, and with different legal status (Vertovec 2007: 1024; see also Crul, Schneider, and Lelie 2013). In a super-diverse city, the fluid, everyday encounters on a local shopping street have a particularly important role to play. They can create public spaces that capitalize on difference but are also inclusive and egalitarian in multiple ways (Hall 2015; Hiebert, Rath, and Vertovec 2015).

In recent years, global migration has brought super-diversity to most of the local shopping streets we write about in this book. East Asians, South Asians, and Central Asians, for example, now own local shops on the same streets in Toronto, Amsterdam, and New York. Customers at a business in Berlin are heard speaking Turkish, Polish, Bulgarian and Russian. Halal meat stores serve observant Muslims from many countries who shop on Fulton Street in Bedford-Stuyvesant, a historically African-American and Caribbean neigh-borhood of Brooklyn, and on Javastraat, a working-class shopping street in Amsterdam originally settled by native-born Dutch (see Table 1).

This degree of super-diversity is increasingly common on local shopping streets in cities around the world, including Paris (Lallement 2010), London (Hall 2015) and Vienna (Heide and Krasny 2010). But different migration paths shape different types of habitus.

Table 1 Local Shopping Streets from New York to Shanghai

City	Street	Location	Description
New York	Orchard Street	Near center	Immigrants to hipsters
	Fulton Street	Near periphery	African American to ethnically diverse
Shanghai	Tianzifang	Near center	Factories to arts district
	Minxinglu	Near periphery	Working class
Amsterdam	Utrechtsestraat	Near center	Upscale but "cozy"
	Javastraat	Near periphery	Immigrants to hipsters
Berlin	Karl-Marx-Straße	Between center and periphery	Immigrants to gentrifying
	Müllerstraße	Between center and periphery	Working class, ethnically diverse
Toronto	Bloordale	Near center	Immigrants to gentrifying
	Mount Dennis	Near periphery	Working class, ethnically diverse
Tokyo	Azabu-Juban	Near center	Upscale but "cozy"
	Shimokitazawa	Near periphery	Hipster businesses

Note: These twelve local shopping streets comprise three geographical types: four fairly short streets, which we studied in their entirety or near-entirety (Orchard Street, Minxinglu, Utrechtsestraat, Javastraat); five small, socially coherent segments of long, linear streets that pass through several different neighborhoods (Fulton Street, Karl-Marx-Straße, Müllerstraße, Bloordale, Mount Dennis); and three self-contained shopping districts (Tianzifang, Azabu-Juban, Shimokitazawa).

Global North/Global South, Upscale/Downscale

Global North migrants often open businesses in upscale and gentrifying local shopping streets. This gives rise to a remarkably consistent look and feel on those streets in vastly different cities of the world. By contrast, migrants from the Global South tend to open businesses in low-income neighborhoods, which creates an equally consistent look and feel, but of a different type.

In Azabu-Juban, for example, an affluent neighborhood in Tokyo, the local shopping street features century-old stores selling hand-made kimonos and traditional *sembei* (rice crackers), as well as a pastry shop selling French-style *macarons* (macaroon cookies), an English tea shop, and a New York-inspired café. Because of these stores, and the visible appearance of the people who shop there, including Westerners who live in the neighborhood, the street looks and feels very much like Utrechtsestraat, an upscale local shopping street in Amsterdam.

On the other hand, it is usually migrants from the Global South with little or no financial capital who open small businesses in the shopping streets of low-income areas. They are mainly attracted by low rent. But they also hope to find customers for the low-price goods they can afford to stock. Especially in times of economic decline—when factories close, nearby residents lose their jobs, and storefronts may be empty—a local shopping street can be sustained by new migrants who cannot afford to open a business elsewhere. This is what happened to our shopping streets in Toronto and Berlin, and to Javastraat in Amsterdam, where migrants from the Global South opened dollar (or 1 euro) stores and grocery stores.

It's a little bit different in Shanghai. Historically, a large number of local shopkeepers in this city have been migrants from other parts of China. Today, less educated migrants from nearby provinces own stores on Minxinglu, a shopping street in a working-class district far from the center of the city that looks like working-class shopping streets anywhere. But in Tianzifang, an upscale arts district near the center, business owners include more sophisticated Chinese from Hong Kong and Taiwan, non-Chinese entrepreneurs from Japan, Australia, the United States, and Europe, and Chinese from the mainland who have traveled overseas. They are translating transnationalism into a local shopping street in a way that looks and feels like gentrifying streets throughout the Global North (see Figure 3).

Figure 3   The ABCs—Art Galleries, Boutiques, and Cafés—Bring Transnational Signs of Gentrification to Tianzifang, in Shanghai

Source: Photo by Sharon Zukin.

Such mixtures and fusions have become a familiar sight in cities around the world—so common, we take them for granted. Yet they make a significant point. For different reasons and with little formal coordination, local shopping streets deliver a highly visible message about the interdependence of social, cultural, and economic processes of globalization and local identity.

A local shopping street brings together in a single urban space a broad array of social, cultural, and economic forces that arise on a global geographical scale (see Table 2). Though transnational migrants may be "pushed" from their home countries by war, poverty, or simply a thwarted ambition to succeed in life, they are "pulled" to different streets by low rents, on the one hand, and previous patterns of migration and settlement, on the other. The goods and services that they offer respond to the tastes of shoppers on that street while contributing, in turn, to the street's overall aesthetic image.

In time, the globalized habitus of the local shopping street becomes the visible face of the neighborhood.

The Visible Face

Researchers often use census data and other official statistics on local residents to describe areas of the city. That data tells us a great deal about who officially resides there—that is, the people who *sleep* there. Yet the men and women who sleep in a neighborhood are not the only ones who use it and for whom the space is important. Indeed, on a shopping street, they are often much less visible, and play less of a role in defining the space, than those who come to the street every day to work and buy. Visitors cannot know who lives inside the houses or apartments, but out on the street and through the plate-glass windows of the stores, they see shopkeepers and shoppers, and what all of them are doing. A local shopping street is the visible face of the neighborhood.

Table 2 Global Sources of Local Identity on a Shopping Street

Global Processes		Local Effects
Economic devalorization/revalorization of land reflecting global capital flows	⟶	Rents
Overseas social conditions: poverty, war, dislocation, ambition	⟶	Immigrants
Global consumer culture, creating tastes expressing social class and other specific identities	⟶	Aesthetics
Networks expressing ties based on ethnic, religious, cultural identities that transcend local boundaries	⟶	Enclaves
		Clusters

From the way it looks, a local shopping street delivers a powerful message about whether a neighborhood is rich or poor, with a majority of one ethnic group or another. This message about the space can be "read" by everyone; it helps to determine who "belongs" there and who, by contrast, is "out of place."

Local shopping streets also present a strong visual message about neighborhood change. When halal meat shops replaced traditional, non-halal butchers on Javastraat, native Dutch residents felt the area was changing around them, and becoming less "Dutch." Later, these residents felt the area was becoming more "Dutch" again when gentrifying bars replaced Muslim-owned shops and cafés (Ernst and Doucet 2014).

Likewise, on Orchard Street, on New York's Lower East Side, when an old Jewish-owned underwear store was replaced by a modern art gallery, and a Bangladeshi immigrant's perfume shop was replaced by a trendy bar, everyone saw the new shops as signs of gentrification. But because Jews and Bangladeshis were not, or were no longer, heavily represented in the surrounding residential population, there was little sense that a specific local identity was being erased. Indeed, the street had already developed a critical mass of the "ABCs of gentrification": art galleries, boutiques, and cafés.

As the ABCs suggest, the aesthetic tastes that are served on a local shopping street make it easy to read which social and ethnic groups "belong" there.

But what about businesses that send mixed signals, like Ali's Trinidad Roti Shop, on Fulton Street, in Brooklyn, which sells takeout sandwiches?

The sandwiches at Ali's roti shop are made with halal meat, but they are wrapped in Indian flat bread, which was brought to the Caribbean by East Indian migrants long ago, and sold for the past twenty years on Fulton Street by Ali, a Muslim man of mixed African and Indian descent. With halal sandwiches, a Trinidad and Tobago flag, and an Afrocentric décor, Ali's is a place of belonging for African-American and Caribbean blacks and Muslims, as well—and everyone else, as long as they have a taste for rotis (see Figure 4).

With a different look and ambiance, local shopping streets show where rich and poor folks "belong." Not only price levels and shoppers, but types of products, how they are displayed, and the way shopkeepers speak with customers, all create the habitus of social class. A local shopping street translates the aesthetic markers of consumers' tastes, from preferences for specific colors to textures and interior design, into cognitive signs of social status (Bourdieu 1984; Small 2004). Who does not know what it means when a café selling espresso made with single-origin, shade-grown, fair-trade coffee beans opens down the block?

Some aesthetic markers of social status appear to be universal on local shopping streets around the world. From Fulton Street to Müllerstraße and Minxinglu, low-price stores are decorated with bright, primary colors, large signs advertising the wares inside, and plastic fixtures. Doors and windows are open to the street, products spill over to sidewalk displays. This look sends a

Figure 4 Ali's Trinidad Roti Shop, in Brooklyn, Delivers a Visual Message of Ethnic, Racial, and Religious Diversity

Source: Photo by Sharon Zukin.

message of "cheap" which may be welcome to many locals, but outsiders describe these streets as messy, dirty, and sometimes even dangerous, depending on whom they see shopping there.

On the other hand, displays in shop windows in affluent Azabu-Juban, on upscale Utrechtsestraat, and in the trendy new boutiques on Orchard Street are carefully arranged. The buildings are old, but the store design is up-to-date, and each shop displays a small, "curated" selection of products. Inside walls are either painted white like an art gallery or expose old bricks like a loft, and shelves and chairs are made of wood, not plastic. At first, these stores stand out because they look so different from the old shops. But gradually, as rents rise and more of the ABCs arrive, the look of the whole street changes.

Whether the look says "upscale" or "downscale" is not just the result of individual store owners' aesthetic taste. The aesthetics of the shopping street is a collective projection of social class and cultural capital.

Nevertheless, local shopping streets are not representations of social difference and cultural capital which have been frozen in time. They are dynamic ecosystems. Moreover, like the city as a whole, they respond to an institutional environment shaped by capital investment and state policies, as well as media images and consumer tastes (Zukin 2010). On these streets, individual lives, and collective choices of aesthetics, geography, and habitus, intersect with the broad structures and institutions of political economy.

A Structural Ecosystem

Local shopping streets tell a story about what sort of place the neighborhood is, and what sort of place it is going to be. Yet this story is shaped by many different "authors," all acting for their own reasons and often fiercely competitive with each other. Sometimes their efforts are coordinated, but often by forces of which the actors are only partially aware.

The three main groups of authors are business owners, building owners, and shoppers. They are, of course, functionally interdependent. Shoppers come to the street because of the local shops, and the stores are there because shopkeepers rent commercial space from building owners.

Shopkeepers selling the same products may cluster together because they make more sales when shoppers can easily go from one store to another, or because the owners are connected through social networks and shared cultures. Most often, however, regardless of how "local" they appear, business owners decide to locate on a specific street for a practical reason: because the rent is affordable.

For their part, building owners may have bought their property or inherited it. They may be shopkeepers in the storefront as well, or they may only be landlords. If the shopkeeper's interest is to keep rent low, the building owner's interest is to maintain or even raise the value of the property, which usually brings higher rents. For this reason, there is always at least a potential conflict of interest between store owners and building owners.

On the other hand, if landlords don't raise rents too high, or if business owners also own their building, they can keep the same shops in place for many years. This can make the ecosystem of a local shopping street both more stable and more socially resilient.

The individual "authors" and their divergent interests are shaped into a coherent space by four very important factors: supply chains that bring products to the stores; demographics of the surrounding residential neighborhood; laws and policies of the state, especially the local state or city government; and media images. These four external factors structure a local shopping street's internal ecosystem (see Figure 5).

Supply Chains

Whether they come from near or far, products travel to local shopping streets along supply chains. The route they travel has an effect on prices, on the cachet or social status of the shops, and on the reputation of the local shopping street as a whole.

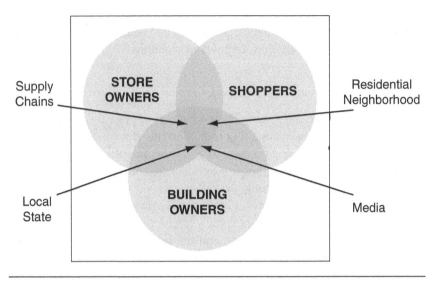

Figure 5   Structural Ecosystem of a Local Shopping Street

A small café on Orchard Street makes espresso with beans imported from Kenya and roasted north of the city, in the Hudson Valley. A boutique on Minxinglu sells inexpensive shoes made—where else?—in China. Each supply chain says something about the character of the street. "Local and artisanal" products that come to the street by bicycle and van have a different cultural valence from goods that are mass produced in a low-wage country, imported by jet plane, and trucked into the city from a regional warehouse.

Local supply chains also work toward environmental sustainability. Few products begin their life in the city. Until urban farms become more widespread, most foods and raw materials will continue to be grown in the countryside. But small-scale artisans and manufacturers are on the rise in many cities, and distribution of their work through local shopping streets has a good effect on environmental sustainability. Notably, moving growers closer to processers and sellers spends less energy derived from fossil fuels. The shorter the geographical distance between links in the supply chain, the greater the contribution to the city's long-term sustainability.

Mass transit lines and bike lanes that run along local shopping streets also contribute to environmental sustainability. They make it easy to shop on the way to, or from, work. Moreover, the density of housing around local shopping streets means that many local shopping streets are walkable and bikeable, because they are close to home. By contrast, ordering online and getting home deliveries from Amazon or Fresh Direct is environmentally costly. When shoppers carry home their packages—or better yet, carry their purchases home in string or canvas bags—they increase the good environmental impact of the street.

For shoppers who value "ethical" consumption, buying locally made products at locally owned shops is important. But for shoppers who are most influenced by price, the length of the supply chain and scale of ownership may be irrelevant. Longer supply chains are basic to supermarkets, discount stores, and fast-food franchises, which all help low-income shoppers to get by.

Residential Neighborhood

We rarely question why low-price stores do business in low-income neighborhoods. Neither do we wonder why kosher meat stores open near a synagogue, nor why halal meat stores cluster near a mosque. Demographics are important on the local scale, as both merchants and shoppers need easy access to their marketplace.

But when new businesses open which look "out of place" on a local shopping street, we wonder whether the neighborhood is changing. Are new shops and restaurants driving residential change, or responding to it?

Probably, it's a little of both. In "gentrification by hipster," for example, young men and women who work in cultural fields move into a neighborhood because the housing costs are low, and when they see there are no businesses to provide a certain kind of good or service which they themselves consume, they decide to open it themselves. We see this on Orchard Street, Javastraat, and Karl-Marx-Straße, and in Shimokitazawa. "There was no place to eat," the new business owners say. "There was no place to hang out and listen to music." These are the same motivations that drive the formation of a critical mass of new business owners in any social or cultural community.

New store owners are *economic* entrepreneurs because they want to make a profit from their business. But they are also *cultural* entrepreneurs because they want to provide a specific product for their own taste community. Moreover, they are *social* entrepreneurs if they aim to make a place where local residents like themselves will feel at home. In all three ways, new businesses re-create the local shopping street, and this fuels a process of change in the residential neighborhood which may have already begun for other reasons.

Yet the same business may be seen differently in, and may have a different impact on, different types of neighborhoods. On Utrechtsestraat, transnational chain stores like Starbucks are looked down on, while on Fulton Street, they are welcome. Why? On the upscale street in Amsterdam, where nearly all shops are owned by individuals and families, chains are a sign of lower social status. But on Fulton Street, where low to middle incomes and ethnic minorities predominate, and until recently crime rates were high, many longtime locals consider the arrival of a moderate-price restaurant chain like Applebee's to be a sign of improvement.

State Regulation

Although local shopping streets seem to be driven by markets, the structural ecosystem as a whole is heavily influenced by the state. All of the ecosystem's central parts—store owners, building owners, shoppers, and supplies—are controlled, or regulated, by policies that are adopted and enforced by government on one territorial scale or another: by the district or city government, the state or province, and the nation-state. Yet governments differ in precisely what they regulate, under which conditions, and how.

Every country makes laws controlling immigration and imports, and many countries enact value-added taxes on purchases that are equivalent to a national sales tax. But for some kinds of regulation, the local level may be more important than the nation-state. In the United States, for example, there is no national sales tax, but many cities and states have income, payroll, and sales taxes. Enforcement of immigration laws also varies in different localities, which can have a major impact on who works in local shopping streets. Underlying it all, the central government's macroeconomic policy on taxes, wages, and money supply affects shoppers' decisions about how much money they have in their pocket to spend.

Whatever the form of regulation, and which level of the state controls it, government always affects business owners' decisions about whether to hire employees, how much to pay them, and how much to charge customers. Though they exert indirect control, all of these laws structure the ecosystem of the street.

On the city level, zoning laws have the most direct impact on local shopping streets because they determine which kinds of businesses can operate legally in specific places, or zones. But not all cities have zoning laws. And their scope is remarkably different in different cities and countries.

In New York, aside from limiting the height and bulk of buildings, zoning laws are quite general. They control whether space can be used for manufacturing, commercial purposes, or residence. They also control the height and density of buildings, whether there should be a plaza open to the public, and, in certain cases, how big the stores can be.

By contrast, in Amsterdam, zoning laws go into much greater detail. They determine the numbers and kinds of business that can legally operate in specific shopping streets and at specific street addresses. These differences reflect not only different political structures, but also different popular understandings of what the role of the state should be: what government can and should be doing.

In Shanghai, we know that the grand scale of urban redevelopment shows the strong, direct role of the city and district governments. But surprisingly, local shopping streets there are shaped by the state's selective *inattention*. In many shopping streets, owners of ground-floor apartments create space for a retail store in their front room, or build space for a store by extending their property into the street. Though these entrepreneurial developments are tech-nically illegal, local government tends to tolerate them for social reasons. Local shops provide employment, especially for laid off workers, pensioners, and migrants. The rent shopkeepers pay provides extra income to working-class landlords, who may want to supplement their pensions or unemployment compensation.

Compared with other countries, the Chinese state *can* exert an enormous influence over land use, as it does in many other spheres of life. Yet it often chooses not to, in favor of commercial development "from below."

Besides zoning laws and development plans, every local state has an arsenal of laws to regulate shopping streets. Building and health inspectors enforce detailed codes for safety and public health, which can impose a heavy financial burden on small shops and restaurants that struggle to comply. Even more important is law enforcement by the police, because that has a huge effect on crime. Businesses that serve alcohol, offer gambling facilities, or support the sex industry are usually subject to a high degree of special regulation.

Another area of state action that has a significant effect on local shopping streets is infrastructure. Opening a new bus line or subway station is good if it brings more shoppers. But merchants on local shopping streets in every city fear street repairs, subway construction, and the widening of roads, because they disrupt traffic, erect barriers, and persuade shoppers to stay away. Likewise, the local state's decision to change the landscaping of a local shopping street may have a serious impact on both aesthetics and logistics. All of these decisions change the image of the street in intended or unintended ways.

Media Images

While the long arm of the local state reaches deep into the operations of a local shopping street, the media create influential images of the street and broadly diffuse them. Not only traditional print and broadcast media, but websites, blogs, and online travel and entertainment guides reach both a local and a global audience.

Individual businesses' websites are designed to appeal to specific groups of customers. In this sense, they reinforce the effect of shop windows and interior display, and expand the store's desired image to a bigger public. But not all retail businesses have websites. Most immigrant-owned, low-price shops and restau-rants don't have them. The owners may be too busy running the business, or too short of capital. Or they just may rely on word-of-mouth promotion by local and co-ethnic customers.

Social media reinforce the image of a successful street and contribute to its "branding." Not surprisingly, trendy boutiques and restaurants, especially those that cater to young adults, get many more tweets, reviews, and "likes." So access to social media produces a digital divide between "have" and "have-not" shopping streets, with the trendy ABCs on one and downscale, often ethnic shops on the other.

When business owners decide to act collectively and form a merchants' association, branding the street becomes one of their first concerns, after sanitation and security. They realize that they face ever more serious competition, not only as individual business owners, but as stakeholders in an urban shopping destination. Their street competes with similar shopping streets in other areas of the city for the attention of mobile shoppers and tourists.

Street branding is fueled by the media. Urban lifestyle magazines and travel websites often list the "best" local shopping streets, and sometimes they give awards for this. Utrechtsestraat, for example, has twice won an award from *Time Out Amsterdam* as the best "*klimaat*" (environmentally sustainable) street in the city. This kind of award goads shopkeepers to develop even more collective branding strategies.

Besides Utrechtsestraat, three other local shopping streets in this book enjoy widespread media attention. Orchard Street is often listed as one of "New York's best shopping streets" (Sorensen 2013). However, in contrast to the old days, when it was the place to shop for bargains, now Orchard Street is praised for new stores offering "vintage flair" and "lifestyle boutique[s]" (Frommer's undated).

In Shanghai, CNN.com lists thirteen shops, galleries, and bars as "the best of Taikang Lu's Tianzifang" (Schmitt 2010). Likewise, the official travel website of the Tokyo Convention and Visitors' Bureau recommends Shimokitazawa for its "streets lined with fashion boutiques, second hand clothes shops and sundry stores," and also its "stylish cafés and bars" and "old Japanese-style bars and eateries" (GoTokyo undated).

By contrast, shopping streets in low-income and working-class neighborhoods rarely appear in online blogs and travel guides, and are certainly not depicted as hip, cool, or trendy. However, their image changes when gentrifying businesses appear and attract media buzz and online reviews. After the Amsterdam city government took Javastraat in hand, for example, and encouraged the opening of new restaurants and cafés, the media began to depict the street as interesting and trendy. Likewise, when public officials in Berlin wanted to revitalize Karl-Marx-Straße, they gave it a new name bound to attract media attention: "Broadway Neukölln."

Global Toolkit of Revitalization

The structural ecosystem of local shopping streets turns out to be quite complex. Although it revolves around three core groups of shopkeepers, building owners, and shoppers, the ecosystem brings together very different institutions and processes, ranging from supply chains and the media, to the local state and surrounding residential neighborhood. Yet a close look reveals similar processes at work in cities around the world. A global toolkit of urban revitalization brings both top-down and bottom-up strategies of entrepreneurialism and gentrification.

Sometimes the results may be surprising. In China, the state plays the predominant role in shaping top-down urban development policies. However, in Shanghai, both the everyday shopping street Minxinglu and the cultural and entertainment zone Tianzifang were created by local entrepreneurs working from the bottom up without government approval. But because of their geographical locations, the results in each place were quite different. In Tianzifang, a central area surrounded by new, upscale development, bottom-up strategies unleashed commercial gentrification, while in Minxinglu, located in a working-class neighborhood far from the city center, building owners just created a "normal" shopping street.

Meanwhile, in Amsterdam, a city that has an unusual reputation for social tolerance and counter-cultural activity, the city government has an equally strong tradition of planning and promoting change. Using their power to choose which businesses can legally operate on shopping streets, the planners have pushed Javastraat toward commercial gentrification. This top-down strategy of revitalization has the effect of displacing immigrant business owners.

By contrast, on Orchard Street, in New York, it's not the state but building owners who have adopted a common strategy not to renew the leases of immigrant-owned stores that sell cheap clothes and leather coats. Instead, these landlords want to attract more ABCs—art galleries, boutiques, and cafés. The local business improvement district (BID) promotes this type of revitalization by organizing an annual block party to celebrate the area's entrepreneurs in "art + fashion," timed to coincide with the city's annual Fashion Week of showings by established design firms. The BID also organizes several food festivals each year to promote local restaurants, as well as art walks featuring the area's growing number of art galleries.

In Toronto, the BIA (business improvement area) in Bloordale also promotes commercial gentrification by branding the street as a place for the ABCs. By contrast, in Mount Dennis, intensive police surveillance aims to prepare the street for revitalization, while making social life difficult on the street for young immigrants, clients of nearby social service agencies, and homeless people.

Different strategies borrowed from the toolkit to promote local shopping streets turn out to yield the same results: a globally recognizable pattern of commercial gentrification. But what looks the same also reveals a more complicated, underlying "structure of common difference" (Wilk 1995), where local actors for their own reasons create a *habitus* for high-status global tastes.

In some shopping streets, the ecosystem is changed by "market forces," namely, a coming together of building owners, new retail entrepreneurs, new residents, and visitors, who form a different market from the one that existed in the past. In other streets, a new market is nudged into existence by the local state through policies that encourage new business development and residential gentrification. In either case, in every city, local actors "package" the local shopping street in remarkably similar ways. Yet they call this globally recognizable package "local character."

The strategies in the global toolkit of revitalization show interconnections between individual actors, capital investment, state regulation, and aesthetic tastes. Among these factors, the local state plays a dramatically significant role. Whether officials give an advantage to some groups of businesses and business owners, or selectively neglect or even actively reject others who have less capital, skills, and political power, the local state determines the ecosystem's right to exist.

Local shopping streets also reflect the increasing activism of new public–private partnerships like BIDs and BIAs. In Toronto, and to a lesser degree, New York and Amsterdam, these organizations have replaced the local state's governance of public space by private-sector, business-oriented, law-and-order management. This makes local shopping streets not only the public face of neighborhoods, but also micro-incubators of the entrepreneurial city and the neoliberal state (Harvey 1989, 2008).

Yet it is important to remember that gentrification usually comes to local shopping streets after decades of economic decline and disinvestment. In U.S. cities, high crime rates and periodic riots by an alienated population discouraged new stores from opening in many low-income neighborhoods from the 1970s to the 1990s. Potential store owners did not want to be robbed, and they knew shoppers would not come if they did not feel safe (Ford and Beveridge 2004). Erecting a crucial barrier to business development, insurance companies would not cover potential damages from arson or riots. Gradually, city governments demolished burned-out and derelict stores along with abandoned housing. Without public or private capital to rebuild, local shopping streets that had once been lively and crowded with shoppers became vacant lots.

Universal Business Improvement Districts

Beginning in the 1980s, local elites in North American cities, most notably New York, have dealt with disinvestment by forming BIDs that take the management of local shopping streets out of the hands—and out of the budget—of the city government. The members of a business improvement district pay a mandatory extra tax to the city government, which is returned to the BID to hire private security guards and sanitation workers who keep the street safe and clean. BIDs also organize street festivals, and install special lighting to celebrate holidays. Without giving up public ownership of shopping streets, the local state transfers broad responsibility for their governance to the BID as a private-sector, business-based, but not-for-profit association.

Business improvement districts follow the initial lead of BIAs in Canada. But since the first BID in the United States was formed in the Union Square area of Lower Manhattan in 1984, this form of organization has circulated to cities around the world, a "policy in motion," with each city adapting it to local institutions and expectations (Zukin 2010; Ward 2011). Rules and memberships vary from place to place; however, in New York, the major actors in BIDs are commercial building owners, an important point which signals the organizations goal of protecting, and even raising, property values. Yet higher property values may have a damaging effect on small business owners who cannot pay higher rents.

Nevertheless, small shopkeepers are BID members, along with big chain stores, as well as the area's major employers, which may include business corporations, universities, hospitals, and museums. All are stakeholders, though in somewhat different ways, in the fortunes of the local shopping street.

New York now has almost 70 BIDs spread throughout the city, in all kinds of neighborhoods (see www.nycbidassociation.org). Both Orchard Street and Fulton Street have BIDs, although the redevelopment vision of each reflects the changing demographics of the surrounding neighborhood. Though the Lower East Side BID, which manages Orchard Street, aims to attract creative professionals and affluent tourists who visit the area's new boutique hotels, the Bedford-Stuyvesant Gateway BID, which manages Fulton Street, tries to balance new businesses that project an image of, and cater to, relatively affluent gentrifiers, and established businesses that cater to a historically black, low- to middle-income population (interviews, 2010, 2011, 2014).

Even if a BID is sensitive to the vulnerability of longtime store owners, its mission does not include saving them from being displaced or shutting down. Laissez-faire for small business has been very much New York City's policy. However, lack of official concern about commercial gentrification stirs disappointment and discontent among both small business owners who are struggling to pay dramatically rising rents and local customers who lament the daily demise of mom-and-pop hardware stores, longtime diners, and even popular, trendy restaurants (for an ongoing record, see the blog Jeremiah's Vanishing New York, http://vanishingnewyork.blogspot.com).

Toronto and Amsterdam have their own forms of BIDs. But neither sets a goal of preserving longtime businesses. Instead, these public–private partnerships take their strategies from the global toolkit of urban revitalization, which often leads to the replacement of traditional, inexpensive stores with new, trendy establishments. Promoting, and even recruiting, these new businesses accelerates the "market forces" that expand the ranks of residential gentrifiers, including creative professionals, students, and more affluent financial investors and corporate executives. Together, the strategies of these public–private partnerships shape a new market for local shopping streets, one which is often hostile to low-price, low-status immigrant business owners.

Whether reshaped by markets or the state, the new habitus of many local shopping streets is remarkably alike. Internet hot points frequented by new immigrants disappear, along with dollar stores and the cheapest shops. New stores sell expensive T-shirts and jeans, and feature cafés with Wi-Fi access, like Starbucks or its local equivalents. In part this new habitus responds to "market forces": higher rents that old businesses cannot afford and changing consumer tastes. But it also reflects the preferences of BIDs, urban planners, and city government as a whole. What stays in the mind is the ironic impression left by the Dutch owner of a new, upscale jeans and sneakers store on Javastraat, in Amsterdam, who named the shop "Div"—for "diversity," he says, because he "likes it" (interview, 2013).

Though it's not easy to transform the longtime social habitus of a local shopping street, it is even more difficult to sustain it in the face of market- or state-led gentrification. Despite praising the "colors" of social and ethnic diversity, no city that we know of directly regulates commercial rents. On the contrary, city governments share a global toolkit of revitalization that pushes local shopping streets to be upscale, cool, or hip. This imperils the sense of "belonging," or moral ownership, painstakingly built on the street by lower-income folks, transnational migrants, and ethnic and racial minorities.

Social Diversity and Moral Ownership

When local shops change from one type to another, longtime residents and users experience a wrenching sense of loss. They have lost their "moral ownership" of the street, a sense of belonging that goes beyond legal property rights, and is based on a deep identification with the culture of the space. Moral ownership derives from patterns of sociability that are learned and reinforced in everyday actions, and become symbols of inclusion. Together, actions and symbols create a sense that certain groups "own" the street—although their sense of inclusion often signifies the exclusion of others.

Moral ownership is most empowering for groups who are excluded from mainstream society and unable to access economic ownership. The author and poet Langston Hughes (1957: 21–2), a Harlem resident in the early twentieth century, expressed the strong sense of social justice in moral ownership of the streets through the words of a character in one of his novels. "I like Harlem because it belongs to me," the character Jesse B. Semple says. "It's so full of Negroes, I feel like I got protection." Though African Americans were often chased out of neighborhoods dominated by whites, in their own neighborhoods they felt unafraid to be themselves. They were not viewed with suspicion on their local shopping street merely because they had dark skin. There, they were completely "at home."

By contrast, the sociologist Andrew Deener (2007) describes recent efforts by store owners on the local shopping street of Venice, in Los Angeles, to drive away homeless men and women as well as residents of an adjacent, low-income African-American and Latino community. New, higher-price stores, a new aesthetic, and events organized by the merchants' association, constructed a trendy habitus where former shoppers and users of the space would likely feel out of place. True, the area had been convulsed by gang violence, and many people were afraid. But the new habitus persuaded the media to recast images of the street and drew investors' interest. Moral ownership was taken by a new group, representing a different social class and identifying with different ethnic and racial solidarities.

Challenges to moral ownership of the streets are as old as urban migration and settlement. But lately, they have become more frequent and more intense. First, the huge scale of transnational migration—and, in China, domestic migration—brings many more strangers from "there" to "here," where they are active authors of urban space. Second, rapid escalation of property values, especially in the biggest, global cities of the world, has unleashed the targeted capital investment in upscale buildings and stores that is loosely called gentrification. These processes activate a conflict of interest between longtime shopkeepers, who are often migrants, building owners, and new business owners who appeal to residential gentrifiers. This conflict of interests often ends in the older stores' displacement.

Globalization and gentrification have dramatically changed the experience of local shopping streets. Many streets are more socially and ethnically diverse than ever before. Others have been homogenized by a hegemonic vision of revitalization that values brand names and chain stores, on the one hand, and hip, cool, and trendy restaurants and shops, on the other. Though local institutions in cities around the world still follow different narrative paths, they often promote a homogenized, glossy vision of the city that puts local identities, and social diversity, at risk.

Looking Forward

Clearly local shopping streets are not just places for shopping. They are cultural ecosystems built from many different individual interests, which nonetheless sustain collective identities. Local shops foster sociability, convenience, and community. They offer the many small business opportunities on which economic and social mobility often depends, especially for migrants. They present a sensual experience of local cultures that makes the city come alive for residents and tourists alike.

No city can survive without calculating the costs of globalization and gentrification, and carefully considering their effects. Cities that value social, as well as environmental, sustainability need a patchwork ecosystem of local shopping streets to support neighborhood economic vitality as well as a broad cultural diversity.

On the environmental front, local shopping streets contribute to every city's goal of being resilient. They are walkable and bikeable marketplaces that offer easy access to, and redundancy of, basic supplies. Just as shoppers should be able to buy everything they need for every day on their local shopping street, so should these streets contribute to decreasing carbon emissions and more efficient use of energy.

Though shopping is universal, each street in this book tells a different story. Enriched by successive waves and forms of globalization, threatened at one time by abandonment and at another by gentrification, local shopping streets reflect the ebb and flow of urban social life, capital investment, and changing demographics. If they can manage to sustain their differences from the mainstream of both central business districts and suburban shopping malls, they will be forces of resistance against the rampant tide of standardization that makes cities look and feel alike.

For all of these reasons, local shopping streets are not only the visible face of a neighborhood; they are vital elements of the city's soul.

The following chapters present a "nested" story of cities and streets, interacting within a sometimes contentious, global-to-local frame. Some of our twelve streets are just a few blocks long and fairly narrow; others are wide thoroughfares with heavy car and bus traffic that run in a straight line for miles. In Shanghai and Tokyo, instead of following a linear grid, three of our four "streets" are really small shopping districts radiating outward from a train or subway station, or spreading through a complex of narrow alleys behind a gateway from the street.

From these differences in urban form, we can already see that local shopping streets are shaped by global toolkits of representation, on the one hand, and local traditions of lived experience, on the other. They are testing grounds for the social theorist Henri Lefebvre's (1991) ideas about how urban spaces are lived, produced, and imagined. What we add to Lefebvre's classic formulation is the interplay between global and local forces, and the influence of the local state. Nowhere is this clearer than in the intersection of shopkeepers' narratives and the biography of the street.

References

Amin, Ash. 2012. *Land of Strangers*. Cambridge: Polity.

Anderson, Elijah. 2011. *The Cosmopolitan Canopy: Race and Civility in Everyday Life*. New York: Norton.

Bourdieu, Pierre. 1984. *Distinction: A Social Critique of the Judgement of Taste*, trans. Richard Nice. Cambridge MA: Harvard University Press.

Bourdieu, Pierre. 1990. *The Logic of Practice*, trans. Richard Nice. Stanford CA: Stanford University Press.

Conforti, Joseph M. 1996. "Ghettos as Tourism Attractions." *Annals of Tourism Research* 23(4): 830–42.

Crul, Maurice, Jens Schneider, and Frans Lelie. 2013. *Super-Diversity: A New Perspective on Integration*. Amsterdam: CASA/VU Press.

Deener, Andrew. 2007. "Commerce as the Structure and Symbol of Neighborhood Life: Reshaping the Meaning of Community in Venice, California." *City and Community* 6(4): 291–314.

Ernst, Olaf and Brian Doucet. 2014. "A Window on the (Changing) Neighbourhood: The Role of Pubs in the Contested Spaces of Gentrification." *Tijdschrift voor economische en sociale geografie* 105(2): 189–205.

Ford, Julie M. and Andrew A. Beveridge. 2004. "'Bad' Neighborhoods, Fast Food, 'Sleazy' Businesses, and Drug Dealers: Relations between the Location of Licit and Illicit Businesses in the Urban Environment." *Journal of Drug Issues* 34(1): 51–76.

Frommer's. Undated. "The Top Shopping Streets & Neighborhoods." Available at www.frommers.com/destinations/new-york-city/663769#sthash.jqLiqnL7.dpbs (accessed June 23, 2014).

Gold, Steven J. 2010. *Store in the 'Hood: A Century of Ethnic Business and Conflict*. Lanham MD: Rowman & Littlefield.

GoTokyo. Undated. "Shimokitazawa." Available at www.gotokyo.org/en/tourists/areas/areamap/shimokitazawa.html (accessed March 14, 2015).

Hall, Suzanne. 2012. *City, Street, and Citizen: The Measure of the Ordinary*. London: Routledge.

Hall, Suzanne M. 2015. "Super-diverse Street: A 'Trans-Ethnography' Across Migrant Localities." *Ethnic and Racial Studies* 38(1): 22–37.

Harvey, David. 1989. "From Managerialism to Entrepreneurialism: The Transformation in Urban Governance in Late Capitalism." *Geografiska Annaler. Series B, Human Geography* 71(1): 3–17.

Harvey, David. 2008. "The Right to the City." *New Left Review* 53: 23–40.

Heide, Angela and Elke Krasny, eds. 2010. *Aufbruch in die Nähe: Wien Lerchenfelder Strasse*. Vienna: Turia + Kant.

Hiebert, Daniel, Jan Rath, and Steven Vertovec. 2015. "Urban Markets and Diversity: Towards a Research Agenda." *Ethnic and Racial Studies* 38(1): 5–21.

Hughes, Langston. 1957. *Simple Stakes a Claim*. New York: Harcourt Brace Jovanovich.

Hum, Tarry. 2014. *Making a Global Immigrant Neighborhood: Brooklyn's Sunset Park*. Philadelphia PA: Temple University Press.

Jacobs, Jane. 1961. *The Death and Life of Great American Cities*. New York: Random House.

Kasinitz, Philip and Bruce Haynes. 1996. "The Fire at Freddy's." *Common Quest* 1(2): 25–35.

Kosta, Ervin. 2014. "The Immigrant Enclave as Theme Park: Culture, Capital, and Urban Change in New York's Little Italies." In *Making Italian America: Consumer Culture and the Production of Ethnic Identities*, edited by Simone Cinotto, pp. 225–43. New York: Fordham University Press.

Lallement, Emmanuelle. 2010. *La ville marchande, enquête à Barbès*. Paris: Téraèdre.

Lee, Jennifer. 2002. *Civility in the City: Blacks, Jews and Koreans in Urban America*. Cambridge MA: Harvard University Press.

Lefebvre, Henri. 1991. *The Production of Space*, trans. Donald Nicholson-Smith. Oxford: Blackwell.

Lin, Jan. 2010. *The Power of Urban Ethnic Places: Cultural Heritage and Community Life*. New York: Routledge.

Lu, Hanchao. 1995. "Away from Nanking Road: Small Stores and Neighborhood Life in Modern Shanghai." *Journal of Asian Studies* 54(1): 93–123.

Min, Pyong Gap. 1996. *Caught in the Middle: Korean Communities in New York and Los Angeles*. Berkeley CA: University of California Press.

Min, Pyong Gap. 2011. *Ethnic Solidarity for Economic Survival: Korean Greengrocers in New York*. New York: Russell Sage Foundation.

Portas, Mary. 2011. *The Portas Review: An Independent Review into the Future of Our High Streets*. December. London: Department for Business, Innovation and Skills. Available at www.gov.uk/government/uploads/system/uploads/attachment_data/file/6292/2081646.pdf (accessed March 5, 2015).

Rath, Jan, ed. 2007. *Tourism, Ethnic Diversity and the City*. New York: Routledge.

Schmitt, Kellie. 2010. "Master Shopper: The Best of Taikang Lu's Tianzifang." April 26. Available at http://travel.cnn.com/shanghai/shop/master-shopper-taikang-lus-tianzifang-166664 (accessed June 23, 2014).

Small, Mario Luis. 2004. *Villa Victoria: The Transformation of Social Capital in a Boston Barrio*. Chicago IL: University of Chicago Press.

Sorensen, AnneLise. 2013. "New York's Best Shopping Streets." April 25. Available at www.newyork.com/articles/neighborhoods/new-yorks-best-shopping-streets-60245 (accessed June 23, 2014).

Taylor, Ian. 2000. "European Ethnoscapes and Urban Development: The Return of Little Italy in 21st Century Manchester." *City* 4(1): 27–42.

Vertovec, Steven. 2007. "Super-diversity and Its Implications." *Ethnic and Racial Studies* 30(6): 1024–54.

Ward, Kevin. 2011. "Policies in Motion and in Place." In *Mobile Urbanism: Cities and Policymaking in the Global Age*, edited by Eugene McCann and Kevin Ward, pp. 71–96. Minneapolis MN: University of Minnesota Press.

White, E. B. 1949. *Here Is New York*. New York: Harper & Bros.

Wilk, Richard. 1995. "Learning to be Local in Belize: Global Systems of Common Difference." In *Worlds Apart: Modernity Through the Prism of the Local*, edited by Daniel Miller, pp. 110–33. London: Routledge.

Zukin, Sharon. 2010. *Naked City: The Death and Life of Authentic Urban Places*. New York: Oxford University Press.

CHAPTER 2

From "Ghetto" to Global

Two Neighborhood Shopping Streets in
New York City

PHILIP KASINITZ AND SHARON ZUKIN

Despite its reputation as a center of global finance and fashion, New York is a city of neighborhoods. New Yorkers are deeply invested in, and protective of, their local communities, and nowhere is their sense of local identity more focused than on the neighborhood shopping street. Yet during the past few years, with rising overseas investment in New York real estate and gentrification hitting ever more dramatic peaks, rent increases have forced many local shops to close. Longstanding, family-owned, "mom-and-pop" grocery stores, hardware stores, clothing shops, and diners have disappeared. Specialty retailers like camera stores have either become obsolete or find it hard to compete with retail websites.

Because of the persistence of old buildings and small storefronts, many local shopping streets *look* roughly the same as they did years ago. Yet the streetscape reveals a growing dominance of businesses selling food and personal and financial services, many of which are branches of local, national, and international chains. Block after block, the patchwork ecosystem is being reshaped by Dunkin' Donuts, Citibank, Walgreen's, and Starbucks, punctuated by individually owned restaurants, hair and nail salons, and espresso bars (Center for an Urban Future 2013).

Nonetheless, local shopping streets remain a vital part of the city's economy and culture. Compared with most U.S. cities, New York retains an old-fashioned resistance to big retail chains. Changes that remade the American shopping experience from the 1950s to the 1990s were slow to arrive. With high land prices and congested streets, New York has been inhospitable to the large-volume stores and discount chains whose business model requires quick access to trucking routes, big parking lots for shoppers' cars, and lots of space for horizontal, single-story, "big box" construction.

The competitive advantages of such stores, and the "one-stop shopping" that they offer, were also less appealing here. Far less likely to own a car than other Americans, New Yorkers are used to making shopping trips on foot, by mass transit, or, in recent years, by bicycle. Moreover, New Yorkers live in smaller homes, with smaller refrigerators. The lack of cars and storage space discourages making biweekly trips to "stock up" at the shopping mall or filling a minivan with a half month's supply of groceries. The city's retail spaces tend to be smaller, too, which encourages store owners to specialize rather than offer a huge selection.

New York is also, historically, a labor union town. Employees, at least in larger stores, including supermarket chains, are often unionized. For a long time this discouraged low-wage, anti-union, national chains from opening in the city.

Finally, many New Yorkers remain proud of what the humorist Calvin Trillin (2013) calls their "ten-stop shopping": shopping "not just for the quality of the goods but for the companionship and the ritual."

Yet all of these factors have not been enough to keep the forces of large-scale retail at bay. Encouraged by the Giuliani administration, which feared a growing loss of sales tax revenues to the suburbs, Kmart opened its first store in Manhattan in 1996. Target soon followed with stores in shopping malls in Brooklyn and Queens. IKEA, facing residents' protests over traffic congestion in one neighborhood, found a more isolated location on the Brooklyn waterfront after promising to hire local residents and pay for improvements in the landscape (Zukin 2010). Only Walmart, the largest of the giant discount chains, has still not opened a store in New York. The chain's labor practices continue to arouse fierce opposition by the city's organized labor movement and elected officials, including the current mayor (Chayes 2014).

By the 1980s, however, many of the legal protections that surrounded local shops had disappeared. "Blue laws" which prohibited sales on Sundays for religious reasons were struck down by the courts. Though many small, independent shopkeepers were reluctant to hire extra employees to work on Sundays, suburban chain stores did so, and they soon lured New Yorkers who owned cars from the city for Sunday shopping. Department stores in the central business districts also began to open on Sunday, which drew shoppers who otherwise would have remained in their neighborhood.

With more women working outside the home, "ten-stop shopping" became less practical. As Trillin (2013) writes about a family-owned, Italian mozzarella maker that closed their tiny retail shop in his now-gentrified, lower Manhattan neighborhood, "A lot of people who now live within walking distance of Joe's

Dairy favor one-stop shopping, because they don't have time for the nine other stops; unlike the [old] Italian women from the tenements, they're in an office all day."

The fact that a traditional mozzarella maker survived until the twenty-first century in modern Manhattan testifies to the pervasive influence of globalization on local shopping streets throughout the city's history. From colonial days, European immigrants and their butcher shops, dairies, and other small stores, dominated New York's retail landscape. Over time, they were joined by many more transnational migrants, who established "ethnic niches" in specific kinds of retail business both inside their own ethnic enclaves, like Chinatown and Little Italy, and in every other kind of community.

With the resumption of large-scale immigration in the late 1960s, newcomers from more regions of the world arrived. By the 1980s, they were beginning to revitalize local shopping streets throughout the city, many of which had been starved for capital investment by years of suburban flight, civil disturbances, and the city's own fiscal crisis. By the early 2000s nearly half of all small business owners in New York were foreign born. Moreover, 90 percent of laundry owners, 84 percent of grocery store owners, 69 percent of restaurant owners, and 63 percent of clothing store owners were immigrants (Fiscal Policy Institute 2011).

At the same time, in gentrifying neighborhoods, traditional, immigrant-owned mom-and-pop stores began to be replaced by art galleries, boutiques, and cafés: the global "ABCs" of gentrification. The small, sometimes quirky spaces of New York's local shopping streets turned out to be well suited to small-scale, artisanal production as well as upscale display of "curated" products. From the East Village to Williamsburg, streets in working-class neighborhoods that had been dominated by low-price stores yielded to gentrification by students and hipsters (Zukin and Kosta 2004).

Today, globalization, immigration, and gentrification are the major forces reshaping local shopping streets in New York City. Yet they create contrasting locales with different kinds of social status and cultural capital. On the one hand, local shopping streets in working-class neighborhoods host mainly low-price stores owned by migrants from the Global South. On the other hand, local shopping streets in rapidly gentrifying neighborhoods have art galleries, boutiques, and cafés, mainly owned by migrants from the Global North.

To illustrate the divergent ecosystems that they form, we focus on two local shopping streets in areas that were historically labeled "ghettos": Orchard Street, on the Lower East Side of Manhattan, and Fulton Street, in the Bedford–Stuyvesant section of Central Brooklyn. Though Orchard Street has experienced a wave of commercial gentrification led by trendy restaurants and upscale "ABCs," Fulton Street has been revitalized by downscale stores, often owned by immigrants. These forms of globalization and gentrification are fed by real estate markets, changing consumers' tastes, and continued migration from both the Global North and Global South (see Figure 1).

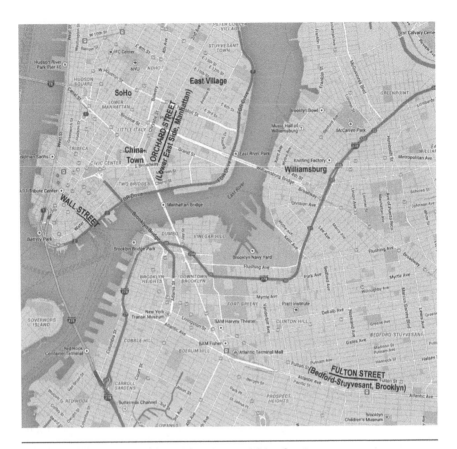

Figure 1 Map of New York, Showing Orchard Street and Fulton Street

Source: Google Maps, adapted by Sebastian Villamizar-Santamaria.

Two Shopping Streets: Orchard Street and Fulton Street

Orchard Street is narrow and only six blocks long, running from Houston Street on the north, near the East Village, to Division Street, on the south, on the border of Chinatown. It is the best-known shopping street on the Lower East Side, a neighborhood settled by successive waves of European, Hispanic, and Asian migrants and, more recently, by "neo-bohemian" artists and creative professionals (Lloyd 2006). Most of the four- to seven-story buildings on the street are tenement houses that were built in the late nineteenth century, with shops on the ground floor. A few buildings also have stores in the basement, which are entered by stairs from the sidewalk, creating a total of around 175 retail businesses. Because of the footprint of the buildings, practically all of the stores are small, and even today only four or five appear to be branches of local chains (see Figure 2).

(A)

(B)

Figure 2 (A) "Old" Orchard Street, 2013. The Upper Floors Have Been Renovated to "Luxury" Rental Apartments, and Today These Stores Are Almost All Gone. (B) "New" Orchard Street, 2014. Espresso Bar, Designer Boutique, New Shoppers

Source: Photos by Sharon Zukin.

Fulton Street is a longer and wider thoroughfare, running across Brooklyn and connecting several of the borough's historically African-American neighborhoods. We focus on a six-block-long section of the street to the east and west of Nostrand Avenue, which is generally regarded as the commercial core of Bedford–Stuyvesant. Most of the two- to four-story buildings on these blocks are houses or apartment houses that were built in the early twentieth century, with shops on the ground floor. Unlike on Orchard Street, Fulton Street has heavy car and bus traffic, and a subway line runs underground. Today there are around 175 businesses on these six blocks, roughly the same number as on Orchard Street. Some are as small as shops on Orchard Street, while others are two to four times that size.

Reflecting the high foot traffic, fast-food chains cluster around the subway stations. However, indicating differences in the investment climate on the Lower East Side and in Bed–Stuy these days, rents for storefronts on Orchard Street are two to ten times higher than on Fulton Street (see Figure 3).

Both streets are located in traditionally poor to working-class neighborhoods, outside of, although a reasonably short subway trip from, the city's central business districts. For a long time, both streets were decidedly downscale shopping areas clearly identified with specific ethnic populations: with

Figure 3 Afternoon on Fulton Street
Source: Photo by Sharon Zukin.

Jews, and later Asians and Latinos, on Orchard Street, and African Americans and Caribbean Americans on Fulton Street. For these reasons, both streets, and both neighborhoods, have often been referred to as "ghettos." Indeed, the Lower East Side is the first American neighborhood to which this term was widely applied, in the early 1900s (Hapgood [1902] 1983). Bedford–Stuyvesant, one of the largest African-American neighborhoods in the United States, has been described as a "black ghetto" since the 1940s (Connolly 1977).

Throughout the twentieth century both streets were known for bargain shopping, in a double sense. Many low-cost goods were available, often displayed on racks on the sidewalk, and, unlike in most stores in the city, shop-keepers and customers would engage in spirited haggling to negotiate prices. As recently as the 1990s, Orchard Street was recommended by online tourist guides as a place where savvy customers who didn't mind the absence of décor and fitting rooms could get a good "deal" for clothing, shoes, and underwear with off-market labels.

Both streets were beset by a sharp rise in crime from the 1970s to the 1990s. On Fulton Street, drug dealers operated openly, even during the day, and busi-nesses shuttered their doors after dark. As in many high-crime areas in U.S. cities, merchants and customers were separated by bulletproof Plexiglas shields. In both cases, the surrounding neighborhood suffered from dilapidated and abandoned buildings and decreasing populations.

Both Orchard and Fulton Streets saw a marked upswing in business from the late 1990s, when crime declined in these neighborhoods, as it did citywide. With improved public safety both areas experienced an influx of more affluent and mostly white residents. Yet both Bed–Stuy and the Lower East Side still have large concentrations of public housing. This guarantees that, despite gentrification, both neighborhoods will be home to residents of different income levels for some time to come.

On both streets local property owners organized through business improve-ment districts (BIDS) have tried to change the image of the street to attract more shoppers. In doing so, both BIDs have shown an ambivalent attitude toward the strong ethnic and cultural identity of the historic local ecosystem. At times the visible remnants of the streets' Jewish and African American iden-tities have been used as a resource for business promotion. But at other times they have been treated more as a burden to be overcome.

Today both streets are in transition. Like the neighborhoods that surround them, both are contested terrains. Yet the city's dominant processes of global-ization, immigration, and gentrification are playing out very differently in the two locales. Orchard Street's transformation from a bargain district of down-market clothing stores to an enclave of "hip" venues of cultural consumption is all but complete. Though Fulton Street is contending with many of the same forces of change, its future seems far less certain.

Close Up: Orchard Street

> Orchard Street. The crush and the stench were enough to suffocate one: dirty children were playing in the street, and perspiring Jews were pushing carts and uttering wild shrieks … Was this the America we had sought (Howe 1976: 67)?

With this harsh description, the literary critic Irving Howe captures the abrasive social life in the teeming commercial landscape of Orchard Street in the first decade of the twentieth century. The street's tenements were overcrowded with extended families and boarders. Residential, commercial, and recreational uses jostled for space. On the street itself, in addition to kosher butcher shops, candy and cigar stores and all kinds of low-price retail businesses, pushcart vendors aggressively sold their wares. Orchard Street epitomized the dirty, bustling, chaotic immigrant Jewish ghetto in the minds of most Americans.

However, as early as the 1920s, upwardly mobile Jewish immigrants and their children began to leave the Lower East Side for middle-class neighborhoods farther from the city center (Moore 1981; Zimmer 2007). Population pressure was also reduced by national immigration restrictions in 1924 (Foner 2000). During the 1930s and 1940s, the area was targeted for modernization by urban planning commissioner Robert Moses and the activist, New Deal Mayor Fiorello LaGuardia. Using federal as well as city funds, they demolished tenements and erected high-rise public housing projects. They also banned pushcarts and street vending, which LaGuardia—himself the child of Italian and Jewish immigrants—saw as unsanitary, outdated, and embarrassingly "old world." In their place the city built indoor public food halls in many immigrant neighborhoods, including the Lower East Side (Bluestone 1992).

The decrease in the residential population, the banning of pushcarts, and the larger economic crisis of the Great Depression caused a shift of businesses on Orchard Street. Most of the food stores left and a concentration of wholesalers, "jobbers," and retail shops at the low end of the garment industry emerged, along with ancillary businesses such as tailors and fabric stores. In the face of declining demand and increasingly stringent health and safety codes, many cash-strapped landlords chose to leave upstairs apartments vacant or converted them to storage spaces for the stores below.

After World War II, migrants from Puerto Rico and later from Latin America and China settled in the area. Though the Jewish residential population continued to decline, Jewish merchants remained the key retail presence on Orchard Street. Their numbers were replenished by Holocaust survivors as well as Jewish immigrants from Latin America and Russia. As many shopkeepers were observant Jews, most stores closed on Saturday and opened on Sunday instead.

Despite this flagrant violation of the blue laws, Sunday shopping became a signature feature of Orchard Street. It was finally legalized in the 1970s by the city government when the northern part of the street was designated a Sunday pedestrian zone and closed to automobile traffic. Sunday shopping, and the pedestrian mall, continued despite the arrival of new, non-Jewish immigrant store owners from South Asia and China in the 1980s and 1990s. Like similar shopping streets in London, Amsterdam, Berlin, and Toronto, the street retained the look and atmosphere of an open-air bazaar.[1]

A Metropolitan Bazaar

Shoppers came to Orchard Street not only from the surrounding neighborhood, but from the entire metropolitan area, searching for bargains. Office workers, bankers, and stock brokers came to shop for suits and designer dresses at deep discounts. Nurses and housekeepers shopped for uniforms, and men who loved elegant shoes made from exotic leathers found them there, too.

The concentration of many similar stores—selling clothes, leather coats and shoes, and fabrics—points to a paradox of this type of small retailing. Traditional Orchard Street merchants were, and the few survivors still are, fiercely competitive. They will go out on the street and pull any passerby looking even remotely interested in an item into the store, while verbally promoting the prices and quality of their shop's goods in contrast to neighboring stores, where apparently identical goods are on offer. Yet it was precisely the concentration of similar stores that attracted shoppers to the street. If shoppers did not find an item they wanted at a good price in one store, they could get it, or another very much like it, nearby. And the merchants of Orchard Street, whether they were old immigrant Jews from Russia or new immigrant Muslims from Pakistan, understood that.

During the 1980s and 1990s, new immigrants from Pakistan and Bangladesh generally opened the same types of stores as their older Jewish predecessors. The "bargain" atmosphere remained the same, and many of the old Jewish families continued to own the buildings. Garment-related uses occupied almost 90 percent of Orchard Street's retail spaces in 1981, and, despite a considerable change in the ethnic composition of the merchants, dropped only slightly by 1990. However, this changed dramatically after 1990, especially in the early 2000s (Figure 4).

Hipsters Arrive

The Lower East Side's low rents have long attracted a small number of artists, musicians, and students. But during the 1980s, when rents rose in adjacent neighborhoods—first, in SoHo and then, the East Village—their numbers increased, and the Lower East Side was "sold" as another bohemia (Zukin [1982] 2014; Mele 2000). Radical artists' collectives thrived on the side streets, just around the corner from the discount stores. But the streets were still gritty and dangerous, and few thought this would open the door to gentrification.

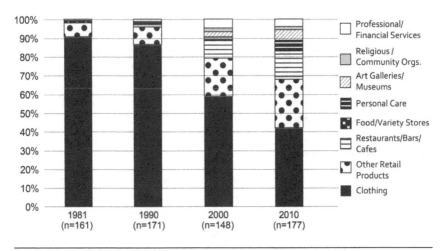

Figure 4 Types of Retail Business on Orchard Street, 1980–2010

Source: Cole's reverse telephone directories; walking census by New York research team.
Graph constructed by Laura Braslow and Sebastian Villamizar-Santamaria.

Yet traditional bargain shopping on Orchard Street was in trouble. During the 1980s, competition from larger discount stores, including the local discount chain Century 21, which also featured deep discounts on designer goods in a no-frills atmosphere, as well as branded stores like Benetton and Gap, all began to erode the street's appeal. As the son of a longtime merchant told us, "Century 21—that's Orchard Street under one roof!"

In addition, an influx of Chinese-owned businesses at the southern end of the street provided goods and services to the growing Chinese ethnic enclave in nearby Chinatown (see Zhou 1992). These included construction firms, restaurant supply stores, and print shops making menus for Chinese restaurants.

As shown in Figure 4, the predominance of garment stores began to wane during the 1990s. Businesses on Orchard Street became more diverse—yet that very diversity threatened the business model of the older stores, whose concentration and reputation for intense competition had made the street a destination. The Pakistani owner of one of the few remaining leather coat stores observes that his customers, mainly blacks and Hispanics from Brooklyn, Queens, New Jersey, and Pennsylvania, still come to Orchard Street to find good deals. However, he is pessimistic about the future. With fewer stores in competition, there is less reason for his customers to come to Orchard Street.

Many of the older shopkeepers suspect that this is precisely what building owners want to happen. As another store owner notes, "My lease is running out. The landlord doesn't want me here anymore. Even though there are empty storefronts, the owners want higher paying tenants." But this merchant has lost customers with the bargain district's decline: "I used to have four stores and now I only have one. People used to come here because it was a bargain shopping district, from out of state, but it's not the same anymore."

For the most part, the older business owners expect that their stores will not be there in a few years. They generally accept this as sad but inevitable, and have neither time nor inclination to mobilize other store owners, who are still their competitors, to protect their collective interests. Nor do they see the BID attracting new bargain seekers. Instead, the BID is looking toward the building of more boutique hotels in the area to bring affluent tourists as customers, and hopes a major mixed-use commercial and residential development a few blocks away will bring middle-class shoppers.

Faced with these prospects, the owners of cut-price stores are just trying to squeeze more time, on a month-to-month basis, from their landlords. "I was served an eviction notice," a leather coat store owner says, "but I'm trying to stay in the store until after the [Christmas] holiday."

History versus Hipsters

At first glance, it doesn't make sense to evict long-established businesses and allow storefronts to remain vacant until new, more attractive businesses appear. Yet in the long run, waiting for the ABCs will establish a new reputation for Orchard Street as hip and trendy, and attract new types of customers. This in turn will permit building owners to raise both commercial rents for the storefronts and residential rents for the apartments above them.

With new businesses beginning to form a "restaurant row," landlords have improbably carved luxury apartments out of ancient tenements. A one-bedroom apartment on Orchard Street rents for at least $2500 a month and a two-bedroom condo sells for almost $2 million, while the rent for a 2500-square-foot storefront with a basement is listed as $28,000 a month. Until recently, this space was occupied by a "manufacturer's outlet" store for leather coats, whose owner may have paid half or only one-third of that rent. Now, a local real estate agency describes it as an "Absolutely Perfect Location for a Massive Restaurant. Perfect for Retail, Nightlife, Flagship [Art] Gallery" (Misrahi Realty undated).

Though the new rents are impossibly high for the owners of small bargain stores, the owners of ABCs rarely complain. In contrast to other trendy areas in Manhattan, such as SoHo and Chelsea, Orchard Street is still a bargain. By the same token, the few, mainly Jewish store owners who bought their buildings years ago feel little pressure to leave.

Yet all around them, Orchard Street has become a cultural destination. In 1992 the Lower East Side Tenement Museum opened in a restored 1860s tenement building toward the north end of the street. The museum, its gift shop, and its walking tours soon attracted middle-class tourists, mostly from around the U.S., local visitors, and school groups. Fifteen years later, the New Museum, an exhibition space for contemporary art, moved to the Bowery, a few blocks away. Art galleries, espresso bars, and ambitious chefs soon followed, encouraged by local real estate agencies.

From the 1970s, property owners and merchants have made efforts to improve Orchard Street's reputation. Since the mid-1990s, however, when the BID was founded, there has been a shift in strategy. Early efforts to promote the street traded on history, nostalgia, and the area's immigrant Jewish identity. Sunday shopping, aged buildings, and colorful bargaining were sold as part of an "authentic" old New York experience, particularly for shoppers returning to what had been their grandparents' neighborhood.

Some recent promotions continue in this vein. An optician's shop that has been on Orchard Street for almost a century displays old black-and-white photographs of the street in their windows and on their website (Moscot undated), and the BID holds an annual Pickle Festival. But the pickles include Korean and artisanal Brooklyn specialties, for kosher dills have not been brined and sold on Orchard Street for years.

The opening of a café by Russ & Daughters, the only surviving Jewish "appetizing store" in the area, shows how Orchard Street's roots can be used successfully to promote upscale commerce. With its high-quality smoked fish, served in a meticulous re-creation of early-twentieth-century delicatessen décor, the café offers the experience of the Lower East Side to a national and even an international clientele of cultural consumers. The *New York Times* named it the second-best restaurant to open in New York City in 2014.

Shopkeepers' Stories

From Immigrant Dream to Hipster Migration

For one hundred years, shopkeepers on Orchard Street spun a succession of narratives about the "immigrant dream." First, Jewish immigrants rose from selling their wares from pushcarts to owning real retail stores; then, Pakistanis, Russians, and Chinese took their place alongside them selling goods and services at bargain prices. "I am from the Dominican Republic," one of four tailors, all Dominicans, who own shops on Orchard Street, says. "I have owned a tailor shop for eleven years, and been located on Orchard Street for seven."

If for several generations Orchard Street represented the immigrant dream of small-scale business ownership and upward social mobility, it now represents the inevitability of these businesses disappearing, and being replaced by market-led commercial gentrification. It is this "hidden hand" of the market, though attached to the landlord's long arm, which persuades the remaining bargain store owners there is nothing they can do to prevent their demise. "Before, there were thirty-five leather coat stores on this block," a remaining store owner says. "Now, there are only six."

In contrast to immigrants who opened businesses on Orchard Street in the 1980s and 1990s, todays new immigrant entrepreneurs come from countries in the Global North. Mainly newcomers from Europe, Japan, and other parts of the U.S., they open hip boutiques and trendy restaurants. They know little of the Lower East Side's history and see little of value in the street's ethnic past.

Like gentrifiers everywhere, these hipster migrants often date the street's revitalization to their own arrival. As the manager of a restaurant–café told us:

> When we first opened [in 2003], it was only Barrio Chino [a popular Mexican restaurant and bar] around the corner, and us. Everything else was lingerie stores … [The space that is now] Café Mezcal was an underwear store … [The other stores] were all low-end and immigrant-owned. They must have made a sale every other day.

Before the manager's own restaurant opened, their storefront "used to be a mattress store. [When the new owner took over,] she completely changed it. She [deliberately] destroyed everything."

An art gallery owner who opened on Orchard Street in 2005 remembers that when she came here,

> the stores all sold clothes … It was a pioneer experience because we were just about the first gallery on the street. In the past two years, four to five galleries have opened just on this block. An art gallery in Chelsea branched out here, and there's a rumor that another is coming. An arts district isn't "emerging"; it has already emerged!

In fact, the first art gallery on the street opened in the 1990s, and the gallerist is the son of the longtime owner of the uniform store. But gentrifiers often like to say they were the first to arrive.

Another typical narrative of, broadly speaking, hipster gentrification is that of a clothing designer and shop owner who became so successful, he opened two more boutiques in Brooklyn. "I moved to New York to pursue a master's degree at the Fashion Institute of Technology," he says. "I own this store and design some of the clothing in a workshop in the backroom." Like him, recent graduates of art and fashion schools who open startups maintain Orchard Street's roots in the city's garment industry. But their wares, and the way they present them, have changed greatly from the old days.

Unlike a no-frills men's hat store that has been doing business on Orchard Street for years, the boutique of a young hat designer offers hats as fashion statements. These are made by hand and carry the designer's label; some are commissioned for the seasonal showings of fashion collections. In contrast to the fluorescent lights and crowded display of items in the old hat shop, the new shop's interior features dim lighting, dark colors, and objects that are arranged in witty or ironic ways. Wooden molds of heads, used for sizing, emphasize the hand-made aesthetic. The shop window shows an artful arrangement of one or two hats and theatrical props, with the designer's name painted in gold letters on the plate glass.

Building Owners and Cultural Cohorts

The narrative the new stores represent is shared by local landlords. No one articulates this more clearly than Sion Misrahi, one of the area's building owners and most active deal makers. Misrahi has deep roots in the historically Jewish Lower East Side. His immigrant parents settled there after World War II with the help of a local Jewish charity. They opened a clothing store on Orchard Street where Sion began selling pants at age fourteen. By twenty-five he owned his own store on the street and began to purchase buildings, making strong connections with his fellow merchants and landlords. By 1991 Misrahi, now in the real estate business full time, was among the founders of the BID. His goal was "rebranding" Orchard Street as a "historic bargain district." "What the South Street Seaport has tried to do we can do in a grittier, non-antiseptic manner," he told the *New York Times* in 1993 (Salkin 2007).

By the early 2000s, however, Misrahi had changed course. "We decided to rent to bars and restaurants who would bring in the hipsters and change the neighborhood," he recalls. He deliberately looked for "hip" night life businesses for his own buildings, and urged other property owners to do the same. "I changed," he told an interviewer in 2007. "Everyone changes" (Salkin 2007). Yet a few years later, he changed course again, helping to replace the initial trendy businesses with boutique hotels and new condominium towers.

For Misrahi and his fellow property owners, taking a chance on hipster merchants and nightlife venues made sense because such businesses transform the image of the neighborhood. Yet this phase may not last forever, a timeline quickened by the relatively short, five-year leases landlords sometimes offer and the dramatic rent increases they demand when the leases end. Or a successful ABC owner might find other locations more attractive, like the gallerist who moved to a larger space in another street nearby and the hat designer who moved his atelier to Paris. Whether the market is soft or the rents are too high, Orchard Street's upscaling has been accompanied by a large number of vacant storefronts.

The street's new customers are often tourists, an increasingly important component of New York's retail economy. The manager of one vintage clothing store reports that her customers are "half tourists, half local." According to the manager of another, "One-third [of our customers] are European or Asian, one-third are designers, and one-third are 'Lower East Side freshie-fresh biddies': female, willowy, long-haired, with a shaggy look." Likewise, the owner of a hip clothing boutique says his customers include visitors from France, Germany, Sweden, and Austria—as well as Orchard Street residents.

A café manager claims that some customers are "old women who have lived close by for decades. [But] we also have young hipsters who have lived here for a year or less." The art galleries' clientele, according to an owner, has also become "more upscale" in recent years. "Justin from England, every time he comes to New York this is his first stop." The manager of another art gallery says, "The owner is from Europe, and he has lots of friends. People come down from [the galleries] in Chelsea; they're closed on Sunday, and Sunday is our biggest day of the week."

Although they are barely aware of the owners of the older businesses, new store and restaurant owners form their own cultural cohort, sharing the same interests, customers, and contacts. Like the earlier cohort of Jewish shopkeepers, they are highly networked. A clothing designer says he buys supplies at two long-established fabric stores: "Zarin's, Belraf's, [and] I get my thread from a girl on Eldridge [Street]. I do things very local." But he socializes with people in his own cultural group: "We all know each other," he says. "Last night I went to a party with the girl from The Dressing Room [a boutique and bar across the street]. There is also a hat shop around the corner … I wear his hats, he wears my clothes."

From Ethnic Ownership to Cultural Cohort

The category of Orchard Street businesses that has increased the fastest has been bars and restaurants. Before 1990, there were practically no eating places on the street. Yet by 2010, the six blocks housed 18 restaurants or cafés and nine bars (Figure 4). With the sole exception of Russ & Daughters café, none of these restaurants offers food or drink that refers to the cultural traditions of the area's old immigrant residents. Neither do they cater to the low-income Chinese, Puerto Ricans, and Dominicans who still make up two-thirds of the Lower East Side's population. The sole Chinese restaurant on Orchard Street features the innovative "Chinese-American" menu of a chef and co-owner who previously cooked at one of the city's highest-rated restaurants.

Yet the new cohort of bars and restaurants on Orchard Street plays a highly visible role in defining the public face of the neighborhood (Ocejo 2014). Again according to the *New York Times*, the number one restaurant to open in New York City in 2012 was Mission Chinese Foods, located in a basement on the street. This was an offshoot of an eccentric "fusion" restaurant in San Francisco owned by a self-taught, Korean-American chef who was raised in Oklahoma: certainly a new way to define the "local" culture.

Though restaurants and bars attract many customers at night, both old and new store owners complain about the lack of daytime foot traffic. The director of the Lower East Side BID speaks wistfully of attracting the offices of creative firms, but other neighborhoods in Manhattan and Brooklyn already compete for the patronage of the "creative class," a factor that, along with the usual financial problems that small retail stores and especially restaurants confront, puts the new ecosystem at risk.

Moreover, the new businesses' dependence on tourists suggests a growing gap between the needs served by ABCs and the needs of longtime, working class residents, who are marginalized by the cultural ecosystem put in place by commercial gentrification. Despite new luxury housing on Orchard Street, poor and working-class residents remain the majority on the Lower East Side, due in large part to the high concentration of public housing.

But even this is changing. More affluent residents are buying cooperative apartments that were built nearby years ago for members of labor unions. Proposals are made to lease land on public housing property to private developers to raise money to renovate the housing projects. With more residents paying market rents, the neighborhood may tip from low-income Asians, whites, and Hispanics to relatively high-income whites.

Is it just by chance that the newest business to open on Orchard Street is a storefront concierge service for apartment owners and visitors who rent space through Airbnb? Taking charge of house keys and packages for paying customers, this business transforms the social role of Jane Jacobs's (1961) local shopkeepers into a commodity.

Of course, shopkeepers need to serve a market. And Orchard Street was never the exclusive "turf" of local residents. From the 1950s to the 1980s, it was a destination for shoppers from throughout the metropolitan area. Neither are globalization and immigration new phenomena on the street. Global production and successive waves of international migration have reshaped it continuously since the late nineteenth century.

What is new about Orchard Street today is the lack of a distinct local character. For all of their unique qualities, most of the trendy boutiques, restaurants, and bars could be in any hipster neighborhood in New York, or indeed in many other cities.

Nevertheless, the BID is going with the flow. Orchard Street and the whole Lower East Side have seen a remarkable growth of new art galleries in the past ten years, and the BID supports an annual gallery walk during Spring Arts Month in May. It also organizes an Art + Fashion night in September, when area designers, boutiques, and galleries join in a block party during the city's Fashion Week. Another annual outdoor festival features dishes sold by local artisanal food producers and fusion restaurants. These events do not bring shoppers to the bargain stores.

The Tenement Museum historicizes the area's ghetto past and ethnic identity as an object of study. Russ & Daughters café draws on it as nostalgia. Yet most of the other new businesses simply ignore it. For a brief moment, Orchard Street is super-diverse. Yet the shift from ethnic ownership to other cultural cohorts puts the street's sense of place, and its "moral ownership," at risk.

Moral Ownership and Upward Social Mobility

Longtime Jewish business owners accept their loss of moral ownership of the street as inevitable. There is less vocal resentment of commercial gentrification than one might expect—and when it does come, it is generally voiced by owners of early ABCs when they face sharp rent increases. The older merchants who also own property in the area are profiting from the transition. They don't regret the exhaustion of the business model of the bazaar. Their own upwardly mobile children are generally not interested in taking over an old-fashioned, low-price store and haggling with customers.

The South Asian, Chinese, and other immigrant merchants who came in the 1980s and 1990s are bitter about recent changes. Yet they, too, feel powerless to oppose them. This resignation contrasts markedly with the anguish over moral ownership that is seen in African American ghettos, as we will see on Fulton Street.

Close Up: Fulton Street

Fulton Street today is the aroma of our kitchen long ago, when the bread was finally in the oven. And it's the sound of reggae and calypso and ska and the newest rage, Soca, erupting from a hundred speakers outside of the record stores. It's Rastas with their hennaed dreadlocks and the impassioned political debates of the rum shops back home brought out into the street corners. It's Jamaican meat patties, brought out and eaten on the run and fast food pulori, a Trinidadian East Indian pancake doused in pepper sauce that is guaranteed to clear your sinuses the moment that you bite into it. Fulton Street is Haitian Creole heard amid any number of highly inventive, musically accented versions of English. And it's faces, an endless procession of faces that are black for the most part—for these are mother Africa's children—but with noticeable admixtures of India, Europe and China, a reflection of the history of the region from which they have come in this most recent phase of the diaspora (Marshall 1985).

Fulton Street is the main commercial thoroughfare of Bedford–Stuyvesant, one of the nation's largest African-American communities. Originally built for middle-class whites in the late nineteenth century, the area saw a huge migration of African Americans from the South and immigrants from the Caribbean between the late 1920s and the 1950s. This influx was accompanied by the near-total exodus of its white population (Connolly 1977; Kasinitz 1992; Wilder 2000).

By the late 1950s, the area was almost entirely black, and was described as New York's second black "ghetto," after Harlem. But Bed–Stuy was home to a mix of social classes. Though some of its stately brownstone houses were divided into rooming houses and small apartments, others supported two or more generations of working- and middle-class black homeowners. Poor African and Caribbean Americans were crowded into the area's tenement buildings and lived above stores on Fulton and other commercial streets, as well as in large public housing projects to the north. This is the cultural ecosystem that nurtured hip hop performers and rappers, including Jay-Z, Mos Def, Lil' Kim, Aaliyah, and Foxy Brown.

In 1958, the noted African-American artist Jacob Lawrence painted "Fulton and Nostrand," depicting the intersection at the center of our research site as a vibrant jumble of pawn shops, grocery stores, bars, a florist's shop, and a tuxedo rental store, amid a chaotic mass of mostly dark-skinned people dressed in bright colors. Yet although the area was rightly seen as a black community, and was a center of civil rights activism in the 1960s, most building owners on Fulton Street were, and still are, white. This has been a source of ongoing tension in the community, and was dramatized by Spike Lee's film *Do the Right Thing* (1989), which was shot in Bed–Stuy.

Starting in the late 1960s, racial tension and rising crime led to disinvestment from the area. As in many traditionally black neighborhoods, the exodus of white business owners and the remaining white residents was followed by the outmigration of many better-off African Americans. Fulton Street remained busy during the day, but by the 1980s it was virtually deserted after dark.

Area residents worked with local and national politicians to stem the tide of increased poverty. The Bedford Stuyvesant Restoration Corporation, the first nonprofit, community development corporation in the U.S., was founded in 1967, and still maintains its headquarters at the eastern end of our research site. Restoration has improved the local quality of life in a variety of ways, including housing renovation and revitalization of commercial spaces. It also collaborates with the Bed–Stuy Gateway BID, which was founded in 2009.

Alongside Restoration, two quite different institutions indicate how Fulton Street's cultural ecosystem has changed since the 1970s. The Slave Theater was a meeting place for political rallies during the 1980s, but has since fallen into disuse and legal conflicts over its ownership. Nearby, a large mosque, the Masjid At-Taqwa, was founded by a U.S.-born, African-American imam, and took a key role in ousting illegal drug dealers at the end of the 1980s. Since then, it has become an influential presence on the street, drawing many new immigrants from Africa and Asia who have opened halal restaurants and stores.

At the same time, the surrounding residential neighborhood is being gentrified. In 1975, a few blocks directly north of Fulton Street were designated a historic landmark district, and the designation was expanded to a larger area in 1996. The high quality of the historic architecture attracted an influx of black, white, and mixed-race professionals, who not only sparked a revival in the local property market, they also unleashed a wave of gentrification (Zukin 2012). Encouraged by declining crime rates, new businesses and restaurants have opened, and small but growing numbers of whites, including professionals, artists, and students, have also arrived. So have real estate investors.

New luxury apartment houses are under construction, and houses regularly sell for one to two million dollars—while Bed–Stuy has more families entering homeless shelters than any other community district in New York City (Institute for Children, Poverty, and Homelessness 2013).

In the blocks surrounding Fulton Street, the white population grew from only 3 percent in 1980 to 16 percent in 2010. The Hispanic population also grew slightly, but, in a striking change, the black population declined from 85 percent to 63 percent. Though blacks are still in the majority, whites are highly visible, especially on the residential side streets. This demographic shift stirs the most surprise among longtime residents when they hear white tourists speaking foreign-accented English, who are drawn to the area by Airbnb.

All of these changes—the arrival of black Muslim immigrants from Africa and white European tourists, improvement of the retail landscape, and rapid gentrification—make the issue of moral ownership more complex.

Revitalization, Retailing, and Race

In contrast to increasingly nocturnal Orchard Street, Fulton Street now is crowded with cars, trucks, buses, and pedestrians during the day but remains mostly empty after shops close at 7 or 8 p.m. Most stores are individually owned, but fast food chains like KFC, McDonald's, and the Caribbean-themed Golden Krust are clustered around the entrance to the Nostrand Avenue subway station, along with branches of banks and of local, inexpensive clothing chains. Nearby, along with takeout shops selling pizza by the slice and Chinese dishes, Soul Food Kitchen, a local chain, operates a takeout restaurant. Applebee's, a mid-range chain usually associated with suburban shopping malls, opened a sit-down restaurant in 2005 in the office and shopping complex owned by Bed–Stuy Restoration, and a "healthy" salad bar opened across the street several years later.

Though the racial and ethnic character of the residential neighborhood has changed since the 1980s, the retail landscape on Fulton Street has remained in some ways the same. Storefronts that were empty in the high crime 1990s are now filled with shops, in some cases with several small enterprises sharing a single storefront. A barber, for example, shares the rent for his space with two other businesses, one that sells CDs and DVDs, and the other, socks, belts, and cell phone cases.[2] Price levels remain low. Compared with Orchard Street, the types of businesses remain as diverse as on most local shopping streets. Indeed, while only a handful of businesses on the street today were there in the 1980s, the distribution of the *kinds* of stores on the street has been remarkably stable (see Figure 5).

It is remarkable that the retail landscape of Fulton Street does not reflect the tastes of the growing numbers of gentrifiers who have settled nearby. While

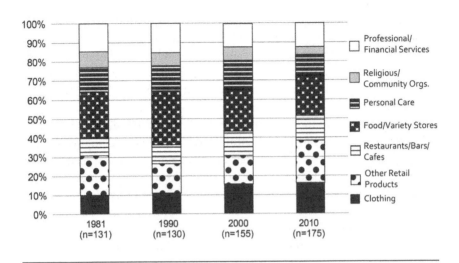

Figure 5 Types of Retail Business on Fulton Street, 1980–2010

Source: Cole's reverse telephone directories; walking census by New York research team. Graph constructed by Laura Braslow and Sebastian Villamizar-Santamaria.

there are upscale cafés, boutiques, restaurants, and bakeries on the side streets, so far these developments have had no visible impact on the area's largest commercial thoroughfare. Perhaps the more upscale business owners feel the narrower side streets have a better vibe. Or perhaps rents on Fulton Street, which are three times higher than on the side streets, represent too great a risk for these types of startups.

Despite a large number of takeout restaurants, Fulton Street has only one supermarket, one health food store, and two or three other stores that sell fresh fish, fruits, and vegetables. However, during the past thirty years, there has been a small increase in clothing and sneakers stores, restaurants and small groceries or bodegas and takeout food shops, and multi-service stores offering overseas money transfers and other financial services. There is also a steady market for shops selling cell phones, religious articles, and wigs.

Many businesses are both owned by, and aim to serve, African-American and Afro-Caribbean customers. In addition to the products that they sell, these businesses sometimes use visible ethnic markers to help establish themselves, and also the street, as a "black public space." Ali's Trinidad Roti Shop, which has been doing business on Fulton Street since the 1990s, features a Trinidadian flag, a painting of an island landscape, and objects on display related to Afrocentric and Muslim themes. Other businesses show the flags of other countries, store names are in various African languages, an occasional poster promotes a candidate in an overseas election. Two surviving record stores are treasure troves of jazz and Caribbean music.

But around the mosque, a significant cluster of halal restaurants, food stores, and African-owned shops has formed. Though African immigrants visibly sustain the racial identity of Fulton Street as a "black" public space, they add a new cultural identity, marking it as both "African" and "Muslim" as well. Moreover, with Muslims from South Asia opening stores and restaurants here, too, Fulton Street can rightly be described as super-diverse (Zukin 2014).

The owners of almost all new businesses near the mosque say they opened there for that reason. Though one halal restaurant owner says his customers are "Jewish, Japanese, [and] whites," it's safe to say that most diners are Muslims, including people who come to the mosque during the day and taxi drivers who eat dinner at the restaurant at night. Likewise, the owner of a halal meat market says his customers include "Caribbeans, Sudanese, [African-]Americans, everybody," but "most" of these are Muslims.

African Americans' attitudes toward the mosque and toward these "African" or "Muslim" businesses are ambivalent. The African-American owner of a café across the street met with opposition from the mosque when he applied for a license to serve alcohol. The aging activists based at the Slave Theater complain that the area's black community is being taken over by "Africans" who, despite their racial commonality, have little interest or investment in Bed–Stuy's historically African-American identity.

Though they are reluctant to express it directly, the attitude toward new immigrants is not all that different from the resentment of Asian and Jewish merchants voiced in other historically black communities (see Kasinitz and Haynes 1996; Gold 2010). At least initially, some Muslim merchants seem aloof from the neighborhood around them: "My customers are mostly my African brothers and sisters" says the man who sells cell phones from a counter in a storefront near the mosque. "I am happy with the flow of fellow Africans and their patronage," says the owner of a money transfer and laptop store.

Others Muslim owners try to express solidarity with their neighbors, such as the Pakistani owner of a halal restaurant who displays a poem he wrote about Martin Luther King Jr. on the wall.

Aesthetics of Low-Price Stores

The aesthetics of new stores on Fulton Street contrast with the look of new shops on Orchard Street. Though some storefronts are as small as those on the Lower East Side, others, especially those of chain stores, are two to four times larger. Many stores and restaurants cover their big, plate glass windows with signs featuring items for sale within and their prices. Some food stores post small signs showing that government food subsidy recipients can use their benefit cards to pay; this sign has disappeared from food stores in many other New York neighborhoods.

Dominant colors on the street are red, yellow, and blue, an almost universal palette for stores in New York's low-income neighborhoods. However, the traditional pan-African combination of red, black, and green is also prominent. Store names are often announced in huge plastic letters, and flashing red and green LED displays promote products and delivery information. When a new store or takeout food shop opens, small, triangular, vinyl flags in red, yellow, and blue are hung on wires around the façade, another sign, at least in New York, of a low-price shopping street.

Unlike many new shops on the side streets, none of the stores on Fulton Street could be called a boutique. In fact, most stores on Fulton Street look like the older shops on Orchard Street. Their interiors are functional, but cluttered. Merchandise is tightly packed on shelves under the harsh light of fluorescent bulbs. In one small clothing store, the carpet is held together by duct tape. And like the traditional bargain stores on Orchard Street, some store owners display inexpensive shirts or dresses on racks on the sidewalk.

As shopkeepers have done on Orchard Street for years, the owners of some small shops on Fulton Street bargain with customers. Within minutes of a customer entering the store, we heard the owner lower the price of a pair of Nike sneakers from $125 to $85, and even offer a payment plan in installments: "Just four payments of $20, and you're done!"

Unlike Orchard Street, Fulton Street offers few places to sit and eat. The one quasi-café where regular customers do seem to be known to the staff is a franchise of the Dunkin' Donuts chain. Customers sit and chat there for hours, and the manager, a man from India, doesn't chase away a woman who asks customers for money or a man who enters selling pirated DVDs. But more upscale cafés do dot the side streets, including an espresso bar twenty steps off Fulton Street that was opened by three young white partners in 2011.

Shopkeepers' Stories

"Not the Ghetto" versus "Still the Ghetto"

Most business owners whom we interviewed on Fulton Street are immigrants. These include West Africans, West Indians, an Egyptian, and a Pakistani, who are all Muslims, but also a Korean, a Russian, an Indian, and a Guyanese, who are not. In contrast, many new businesses owners on the side streets are also largely immigrants, but of a different sort. There are a Parisian pastry maker of African descent and a Haitian-born, former corporate lawyer who now owns a café and a bar, as well as several white chefs and restaurateurs, including two from Italy. Though many of the immigrant store owners have been there since the 1980s and 1990s, gentrifying businesses on the side streets have rapidly increased since 2005.

Regardless of where they came from or how long they have been on Fulton Street, all the business owners tell dramatic stories about the area's history of high crime rates and illegal drug dealing. "It was a wild and bad neighborhood," says the Indian owner of a clothing store. He describes people drinking alcohol and smoking marijuana in the street. They would break locks on the doors and steal from the stores, sometimes taking whole racks of clothes from his shop when his back was turned. "People used to be scared to walk the streets," the Caribbean owner of a health food store recalls. "You never saw a Caucasian in this neighborhood."

As in the rest of New York City, crime rates began to fall sharply in the 1990s and have been falling steadily ever since. "Twenty-five years ago, there were twenty crack houses on the block," says the owner of a halal restaurant, perhaps with slight exaggeration. "There was a 95 percent chance a foreigner would be robbed walking from Bedford to Franklin, while local people would know where the buyers were and where to walk. Today it's the safest block in New York City."

Like most other Muslim business owners, this man credits the "forty days and forty nights" campaign by members of the mosque in 1988 with intimidating the drug dealers so they stopped doing business, at least openly, in the area. But non-Muslim shopkeepers tend to credit the harsh policing and even harsher anti-crime rhetoric of the Giuliani administration. In any case, policing became more vigilant, for which store owners—and longtime residents—are grateful.

Todays safer Fulton Street attracts many shoppers during the day. Yet the area's new white residents are mostly absent. Though on Orchard Street commercial life is *more* gentrified than the residential neighborhood around it, on Fulton Street the retail sector is far *less* so.

A young, white resident who works in the fashion industry moved to Bed–Stuy because of cheap rent and the convenient subway line. He describes the building where he and his roommates live as "a crazy mix, mostly older African-American professionals and [white] kids like us." Yet though he takes the subway from Fulton Street every day, he rarely shops there. He occasionally patronizes new businesses on the side streets that "are not ghetto-looking" and "do not look cheap." But mostly he prefers to shop and eat in more gentrified neighborhoods.

Despite rapid residential gentrification, the persistence of a "ghetto" image on the shopping street is not limited to whites. The West Indian manager of a low-price women's clothing store says that Fulton Street is "too ghetto, [there are] too many junkies." Customers from the neighborhood are "ghetto-like, rough, difficult to argue with."

But that store has gone out of business since we spoke with the manager in 2010. Perhaps because the store owner also owns the building, and sees both the improvement in, and potential for still more upgrading of, the retail landscape, the store is now for rent for $28,000 a month.

Though the vast majority of shoppers on Fulton Street are still African

Americans and Caribbeans, the shopkeepers are well aware of rising rents and the growing white population. According to the African owner of the multi-service store, "My customers are slowly changing as different people come in to use the Internet." The African barber says his customers now "come from all over—the Bronx, Manhattan—and even some whites who have moved into the neighborhood."

A Caribbean takeout food shop owner sees "Caucasian, Manhattan people" moving in, and points to the opening of "more classic shops" and chains like Foot Locker. The Russian owner of a hardware store says the whole neighborhood has changed; "white people come in—can I say that?" The manager of Dunkin' Donuts states that tourists come in, beginning with "French people."

Yet a clothing store owner speaks for many similar stores when he says that his customers are "100 percent black." Most, he reports, live in the neighborhood, and probably half are Caribbean. He finds the new "European" (i.e. white) residents are reluctant to shop in low-price stores like his: "They don't like the style." According to another clothing store owner, the new, young white residents "only shop in the big [department] stores, like Macy's; they only sleep in this area."

Roots versus Empathy

Immigrant business owners have indisputably put down roots on Fulton Street. But because of their different ethnic and religious backgrounds, they form a cultural cohort that differs from the groups with strong black and Christian identities in traditional African-American neighborhoods. Contrasts are both audible and visual. When you walk around on Sunday morning, you hear gospel singing coming from a church and see well-dressed black folks going out to brunch after attending religious services. But you also see brightly painted domes on minaret-like towers decorating a laundromat's façade, and at certain hours of the day, especially on Friday, you hear the call to prayer at the mosque.

The Pakistani owner of a halal restaurant, who wrote the poem about Martin Luther King Jr. and taped it on a wall, feels a deep empathy with the struggles of African Americans. The poem speaks not only about Dr. King, but about discrimination and the police. He told us that he opened his own business because he is "independent minded." The people whom he had worked for abused and even fired him, so he decided to become his own boss. "I would rather eat salt and bread in freedom than eat like a king while being harassed." Yet he also opened the restaurant because he "wanted to do something for the mosque."

It is this "dual consciousness" that anguishes the few black activists who sit in the Slave Theater during the day waiting for passersby to harangue. Though their Afrocentric rhetoric is more exaggerated than the language used by black residents and business owners, they express a shared fear that Fulton Street, and the surrounding neighborhood, are losing their historically black identity. Their fear is not as exaggerated as their rhetoric. When the white-owned espresso bar opened nearby a few years ago, the owners decided to create a "brand" identity linked to the neighborhood's cultural heritage. They displayed old black-and-white photos of streets and houses, which, for them, represent the historic Bed–Stuy.

If most shopkeepers make few efforts to attract new white residents, it is nonetheless clear that both they and their black customers suspect the recent influx of whites will change the neighborhood, Fulton Street included. Unlike on Orchard Street, on Fulton Street our white and sometimes our African-American interviewers were often greeted with suspicion, not only from store owners, but also by passersby. People would stop and ask them who they were, what they were doing there, why they were interested in the local businesses, and whom they were working for. Though few were actually rude, it was clear that young whites, taking notes and asking questions in this part of the city where not long ago white faces were rarely seen, are perceived as a harbinger of rising rents and evictions, as well as a loss of moral ownership.

Moral Ownership and Racial Identity

It is not hard to understand these suspicions. African Americans have historically been far more spatially segregated than other ethnic groups, including those who created Little Italys and Chinatowns (Massey and Denton 1997). They have often been formally and informally excluded from many "public" spaces of the city. So the shopping streets of black neighborhoods came to have a special importance. The "ghetto," while a form of spatial exclusion, could also be a haven, a "place of our own" for those whose presence and patronage were unwelcome elsewhere.

Yet businesses in historically African-American communities have often been owned by outsiders, in many cases "middle-man minorities" such as Jews and later Koreans (Gold 2010; Kasinitz and Haynes 1996; Lee 2001; Min 1996). From Booker T. Washington and Marcus Garvey to Malcolm X and Louis Farrakhan, advocates of greater African-American autonomy and self-determination long called for greater control over the highly visible retail businesses "in our own neighborhoods."

However, just as immigrants tend to be over-represented in self-employment, native African Americans have been unrepresented, particularly in the ownership of retail stores. While self-employment among Caribbean immigrants has been somewhat higher, it tends to be in the professions, not small business (Kasinitz 1988).

The reasons for relatively low numbers of black store owners are both social and financial: racial discrimination, lack of capital, and inability to access credit.

Moreover, since the 1960s, with the expansion of the educated black middle class, the low rates of black small business ownership may reflect a preference for the higher wages and greater job security of white collar employment, especially in the public sector. On the other hand, the greater representation of immigrants among the ranks of store owners reflects their structural limitations: lack of English language skills, U.S. educational credentials (even among those with higher educations in their home countries), and connections to the local labor market outside of their ethnic networks.

But there is a gap between immigrant ownership of many of the stores on Fulton Street and "black" moral ownership of the public space. Though immigrant and Muslim shopkeepers make up a significant presence on the street, African Americans and Caribbean Americans exert the strongest claim to call the street theirs. These groups, along with several whites, constitute the governing body of the Bed–Stuy Gateway BID.

In contrast to the Lower East Side BID on Orchard Street, which is dominated by building owners, Gateway's leaders are professionals closely tied to local government and African-American community institutions. They are at least as answerable to the larger African-American and Caribbean community—particularly its middle class and small business owners—as they are to building owners, who are mostly white outsiders. As the BID's website says, "The Gateway's identity is uniquely punctuated with African American, African, and Caribbean influences" (BedStuy Gateway undated).

As on Orchard Street, the BID leadership envisions "upscaling" the street and revalorizing the surrounding neighborhood. Yet they are also aware that for many in the community, "upscaling" looks like catering to whites and risks a symbolic loss of moral ownership. Therefore, far more than on Orchard Street, the BID must walk a narrow line. They hope to put forward a vision of the street that is middle class, prosperous, less "ghetto"—with "fewer cell phone stores, fewer hair braiding places" —but still identifiably African American. Even if the businesses are downscale, they recognize that their owners represent the longtime ethnic roots of the community.

Yet Gateway's website pays respect to Fulton Street's cultural diversity:

> At night, you can hear the masjid's call to prayer and in the morning, church bells echo off of historic brownstones and pre-war apartment buildings. Cars cruise by pumping out the latest soca, funk, dance hall, and hip-hop, and your neighbors still ask how your parents are doing (BedStuy Gateway undated).

This suggests that business development on Fulton Street can only succeed by promoting a taste for both a close-knit social community and super-diversity.

Yet there is little interaction between the various cultural cohorts of merchants. Moreover, neither the BID nor the store owners are close to the commercial building owners. Despite Gateway's invoking the value of diversity, these groups do not coordinate their activities.

Whose Street Is It?

Orchard and Fulton Streets, like hundreds of neighborhood shopping streets throughout New York City, show how the identities of urban places are continuously transformed by the everyday actions of local building owners, merchants, and their customers. The streetscapes shape New Yorkers' experience of globalization and, increasingly, of gentrification as well. Forms of cultural solidarity, like religion, ethnicity, and "hipster" identity, unite some merchants and customers and exclude others. And both the overlapping and succession of cultural cohorts reshapes the shopping street's ecosystem.

Both Orchard Street and Fulton Street have been revitalized by new investment, restaurants, and retail stores. But Orchard Street has been more successful in shedding its "ghetto" image and transforming itself for the style of consumption identified with the Global North. Gentrification by hipster? Perhaps. But Orchard Street is also adjacent to very expensive districts of Manhattan. Building owners have made a concerted effort to attract the ABCs. The Tenement Museum and now the Russ & Daughters café selectively promote nostalgia for the distant immigrant past.

Though it is too soon to see if these revitalization strategies will be sustained by more upscale development, the new business model has not sparked a crisis over loss of moral ownership. True, a few fifth-generation owners are still running the family store. But Orchard Street's old Jewish identity had already faded: first, in the 1960s, when the residential population changed to mostly Asian and Hispanic, and then, after 1980, when the large concentration of bargain garment and fabric stores began to decline. "Old" immigrants' moral ownership of the street lasted thirty more years, with the support of new immigrants from South Asia and Russia.

The cultural cohort of Pakistanis and Bangladeshis did not contest older business owners' claims to represent the street. To the degree that moral ownership has been an issue, it appears in opposition by Hispanic and Asian residents to the museum's expansion plans: in other words, opposition to the threat of residential eviction.

On Fulton Street the potential loss of moral ownership by longtime shopkeepers and residents is far more complicated. For many, it is frightening. African Americans have fewer places in the city to call their own, and the loss of public space long seen as "theirs" is deeply disturbing. The eastward expansion of gentrification through Central Brooklyn, and the recent redevelopment of Fulton Mall, which was an even larger "black" shopping space in downtown Brooklyn, point to a troubling trend of both real and symbolic eviction. Many longtime residents and shoppers, especially those on meager incomes, hope that

Fulton Street can remain a "black" space because that offers a home to them.

At the same time, local residents are pleased by the decline in crime and the improved business climate. Many, while wanting to see Fulton Street remain "black," wish it were less "ghetto"—that is, less poor—in appearance and shopping options. This desire is shared by blacks and whites, middle-class homeowners, and student and "creative" renters. Indeed, whites often rent apartments from black homeowners, who benefit from rising property values.

As on Orchard Street, but in a different way, the historic base of moral ownership on Fulton Street is changing. Immigration has produced a super-diverse cultural ecosystem. But super-diversity has created a more complicated "black" identity. Like the cultural cohort of ABC owners on Orchard Street, African immigrants and other Muslim business owners on Fulton Street have little nostalgia for the "old" Bed–Stuy. The only business to try to trade in nostalgia is the espresso bar on a side street that is owned by whites.

Throughout the city, local merchants, in the course of trying to make a living, shape the public face of the community. Whether they work in coordination with the local state in BIDs, or individually, on their own, they recreate the patchwork ecosystem of local shopping streets in an intimate way that chain stores and suburban shopping malls do not. Nonetheless, in a time of booming real estate markets, dramatic rent increases threaten their survival. The same story plays out in quite a different city, Shanghai.

Acknowledgements

Working with a team of students from the PhD program in sociology of the City University of New York in 2010–11, we interviewed store owners and managers on Orchard Street and Fulton Street, made ethnographic observations on both streets, and coded and analyzed information on businesses collected from reverse telephone directories and websites. Several of us continued to follow the two streets until 2014, carrying out more interviews and meeting with the directors of the BID on both streets. We thank the business owners and managers, current and former BID directors Michael Lambert, Michael Rafferty, and Bob Zuckerman, and local residents who generously gave us their time, and we are grateful for the hard work and valuable insights of the New York team: Laura Braslow, Benjamin Haber, Jacob Lederman, Sara Martucci, Greg Narr, Julia Nast, Vanessa Paul, Samantha Saghera, Tommy Wu, and Fang Xu. We also thank students in the senior seminar and an MA seminar at Brooklyn College who carried out interviews in Bedford–Stuyvesant from 2009 to 2012.

Notes

1   In the East End of London, the custom of Sunday street markets, originated by Jewish merchants, also continues despite the fact that most vendors now are Muslim.
2   Before the 1990s, these goods would surely have been sold on the street, but crackdowns on street vendors by the police at that time drove them away, or indoors.

References

BedStuy Gateway. Undated. Available at http://bedstuybid.org/ (accessed January 6, 2015).

Bluestone, Daniel. 1992. "The Pushcart Evil." In *The Landscape of Modernity*, edited by David Ward and Olivier Zunz, pp. 287–312. New York: Russell Sage Foundation.

Center for an Urban Future. 2013. *State of the Chains, 2013*. New York: Center for an Urban Future. Available at https://nycfuture.org/research/publications/state-of-the-chains-2013 (accessed December 13, 2014).

Chayes, Matthew. 2014. "De Blasio: Walmart Unwelcome in New York City." *New York Newsday* (June 5).

Connolly, Harold X. 1977. *A Ghetto Grows in Brooklyn*. New York: New York University Press.

Fiscal Policy Institute. 2011. *Immigrant Small Businesses in New York City*. New York: Fiscal Policy Institute. Available at www.fiscalpolicy.org/FPI_ImmigrantSmallBusinessesNYC_20111003.pdf (accessed December 13, 2014).

Foner, Nancy. 2000. *From Ellis Island to JFK: New York's Two Great Waves of Immigration*. New Haven CT: Yale University Press.

Gold, Steven J. 2010. *Store in the 'Hood; A Century of Ethnic Business and Conflict*. Lanham MD: Rowman & Littlefield Publishers.

Hapgood, Hutchins. [1902] 1983. *The Spirit of the Ghetto: Studies of the Jewish Quarter in New York*. Cambridge MA: Belknap Press.

Howe, Irving. 1976. *World of Our Fathers: The Journey of the East European Jews to America and the Life They Found and Made*. New York: Harcourt Brace Jovanovich.

Institute for Children, Poverty, and Homelessness. 2013. *A Theory of Poverty Destabilization: Why Low-Income Families Become Homeless in New York City*. June. Available at www.icphusa.org/filelibrary/ICPH_policybrief_ATheoryofPovertyDestabilization.pdf (accessed January 4, 2015).

Jacobs, Jane. 1961. *The Death and Life of Great American Cities*. New York: Random House.

Kasinitz, Philip. 1988. "From Ghetto Elite to the Service Sector: A Comparison of Two Cohorts of West Indian Immigrants to New York City." *Ethnic Groups* 7(3): 173–203.

Kasinitz, Philip. 1992. *Caribbean New York: Black Immigrants and the Politics of Race*. Ithaca NY: Cornell University Press.

Kasinitz, Philip and Bruce Haynes. 1996. "The Fire at Freddy's." *Common Quest* 1(2) (Fall): 25–35.

Lee, Jennifer. 2002. *Civility in the City: Blacks, Jews and Koreans in Urban America*. Cambridge MA: Harvard University Press.

Lloyd, Richard. 2006. *Neo-Bohemia: Art and Commerce in the Postindustrial City*. New York: Routledge.

Marshall, Paule. 1985. "Rising Islanders of Bed–Stuy." *New York Times Magazine* (November 3): 67, 78, 80–82.

Massey, Douglas and Nancy Denton. 1997. *American Apartheid: Segregation and the Making of the Underclass*. Cambridge MA: Harvard University Press.

Mele, Christopher. 2000. *Selling the Lower East Side: Culture, Real Estate, and Resistance in New York City*. Minneapolis MN: University of Minnesota Press.

Min, Pyong Gap. 1996. *Caught in the Middle: Korean Communities in New York and Los Angeles*. Berkeley CA: University of California Press

Misrahi Realty. Undated. "Famous Beckenstein Building." Available at www.misrahirealty.com/index.cfm?page=details&id=1793 (accessed January 2, 2015).

Moore, Deborah Dash. 1981. *At Home in America: Second Generation New York Jews*. New York: Columbia University Press.

Moscot. Undated. "Our Story." Available at www.moscot.com/Our_Story.html (accessed January 2, 2015).

Ocejo, Richard. 2014. *Upscaling Downtown: From Bowery Saloons to Cocktail Bars in New York City*. Princeton NJ: Princeton University Press.

Salkin, Allen. 2007. "The Lower East Side is Under a Groove." *New York Times* (June 3).

Trillin, Calvin. 2013. "Mozzarella Story: A Cheese Ritual." *The New Yorker* (December 2).

Wilder, Craig. 2000. *A Covenant with Color: Race and Social Power in Brooklyn*. New York: Columbia University Press.

Zhou, Min. 1992. *Chinatown: Portrait of an Ethnic Enclave*. Philadelphia PA: Temple University Press.

Zimmer Amy. 2007. "From Peddlers to Panini: The Anatomy of Orchard Street." In *The Suburbanization of New York*, edited by Jerilou Hammett and Kingsley Hammett, pp. 53–62. New York: Princeton Architectural Press.

Zukin, Sharon. 2010. *Naked City: The Death and Life of Authentic Urban Places*. New York: Oxford University Press.

Zukin, Sharon. 2012. "The Spike Lee Effect: Re-imagining the Ghetto for Cultural Consumption." In *The Ghetto: Contemporary Global Issues and Controversies*, edited by Ray Hutchison and Bruce D. Haynes, pp. 137–57. Boulder CO: Westview.

Zukin, Sharon. [1982] 2014. *Loft Living: Culture and Capital in Urban Change*, 3rd edition. New Brunswick, NJ: Rutgers University Press (first edition: Johns Hopkins University Press, 1982).

Zukin, Sharon. 2014. "Restaurants as 'Post Racial' Spaces: Soul Food and Symbolic Eviction in Bedford–Stuyvesant (Brooklyn)." *Ethnologie française* 44(1): 135–47.

Zukin, Sharon and Ervin Kosta. 2004. "Bourdieu Off-Broadway: Managing Distinction on a Shopping Block in the East Village." *City and Community* 3: 101–14.

CHAPTER 3

Commercial Development from Below

The Resilience of Local Shops in Shanghai

HAI YU, XIANGMING CHEN,
AND XIAOHUA ZHONG

Shopping in Shanghai today is global in many ways. This is obvious to high-end shoppers and overseas tourists when they see brand-name stores like Louis Vuitton and Gucci on Nanjing Road, which may be called the "Fifth Avenue of Shanghai," and an Emilio Pucci boutique on the Bund, the historic waterfront street of the central financial district. But in this chapter we look past the city's glamour zones and instead focus on new and traditional local shops in other neighborhoods that respond in a different way to globalization, migration, and state policies. As always, these three factors are interconnected. Not only is Shanghai deeply penetrated by global markets, its emergence as a global city is driven by a strong state and massive migration from other regions of China.

The overwhelming majority of local shops, restaurants, and other service establishments in Shanghai are owned and operated by millions of migrants from its neighboring provinces as well as more remote areas. They sell cigarettes, fruits, and clothes, operate small eateries and barber shops, remodel people's apartments, and deliver purchases to the doorsteps of local residents. Local shops are the employment "haven" for the majority of almost ten million migrants in Shanghai, many of whom lack the *hukou* documentation, or household registration, of permanent residents. They eke out a living by catering to the daily needs of the city's approximately 15 million residents. Very few local shopkeepers are, in fact, native Shanghainese.

The coupled influence of globalization and migration on local shopping is not confined to China's post-1980 economic reforms and opening to external markets, but dates back to Shanghai's first emergence as China's most cosmopolitan city in the 1920s. That development reflects Shanghai's even earlier status as a "treaty port" connected to Britain and France, beginning in 1842. The International Concessions, which were basically colonial implants of Britain, France, and the United States in the city's central districts and on the Bund, hosted the Western stores, restaurants, theaters, and bars that defined Shanghai's then glamorous character. These foreign-owned oases of prosperity were dazzling, but completely separate from the myriad bustling neighborhood shops that sustained the everyday life of ordinary residents.

Shanghai's encounter with globalization in the first half of the twentieth century shaped its identity and functions as a primarily industrial and banking center. These economic activities were a parallel universe that often overshadowed the lively local commerce represented by widespread neighborhood shops selling cigarettes, grain, coal, groceries, cotton fabric, snacks, fruits, and other necessary items (Lu 1995). One type of these shops—the so-called "Tiger Stoves"—simply boiled and sold hot water for making tea and in-home bathing for a small number of wealthy households with indoor bathtubs. The local shops co-existed with other small private businesses such as hotels and public bath-houses, not only on neighborhood streets but also embedded in the alleyway houses that represented Shanghai's traditional vernacular architecture.

From only about 200,000 residents in 1842, Shanghai's population grew to around five million by 1949, at the dawn of the People's Republic. Most of the new residents were poor migrants fleeing rural poverty and war elsewhere. Besides the draw of factory jobs, many migrants gravitated toward the opportunity to open small shops, restaurants, and service businesses. Shanghai could not have accommodated the large influx of migrants had it not developed a commercial infrastructure of numerous local shops. This historical path of "commercial development from below" laid a foundation for the resilient ecosystem of local shopping streets today.

To understand the resurgence of commercial development from below, we present a parallel case study of two very different shopping streets in Shanghai. Although both are influenced by new globalization from the outside, continuous rural–urban migration within China, and state policies of global city-style urban renewal, each shopping street has emerged in response to specific local conditions.

Two Shopping Streets: Tianzifang and Minxinglu

The two shopping streets we studied are Tianzifang and Minxinglu, located in two very different parts of Shanghai (see Figure 1).

The first street, Tianzifang, is located in an old part of central Shanghai near

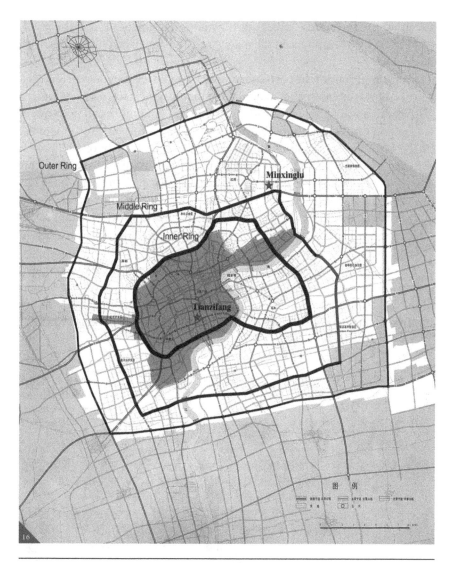

Figure 1 Map of Shanghai, Showing Locations of Tianzifang and Minxinglu
Source: Drawn by Xiaohua Zhong.

the old French Concession area. It did not exist as a shopping street before the early 2000s. However, beginning as an alleyway neighborhood of small workshops and three- to four-story apartment houses about two decades ago, Tianzifang today occupies a number of interconnected long or short narrow lanes filled with art galleries and studios, trendy restaurants and cafés, and boutiques selling clothing and decorative objects. While this spatial layout makes Tianzifang technically a shopping zone as opposed to a linear shopping street, we use the words "street" and "zone" interchangeably, and the area is also known by its entrance on the main street, Taikanglu or Taikang Road (see Figure 2).

Tianzifang is one of the best-known shopping and leisure zones in Shanghai and also boasts an international reputation that attracts many overseas tourists. The area as a whole has earned 4½ out of 5 stars (and more than a thousand reviews) on the travel website TripAdvisor.com and is regularly recommended by the media as a tourist attraction. However, the majority of shoppers in Tianzifang are domestic visitors from the rest of China and Shanghai residents. The regionally diverse but generally affluent visitors stroll the narrow lanes to browse in the stores, drink coffee or cocktails, and dine at fine restaurants. Yet despite its commercial success—and despite the local state's important role in urban development—Tianzifang developed outside governmental channels, in an unplanned and unexpected way.

The second street is Minxinglu (Minxing Road), located in the Zhongyuan area in northeastern Shanghai, about a 30-minute subway ride from the center of the city. Zhongyuan was built up as a large residential district of low- and mid-rise apartment buildings in the 1980s to house the large number of

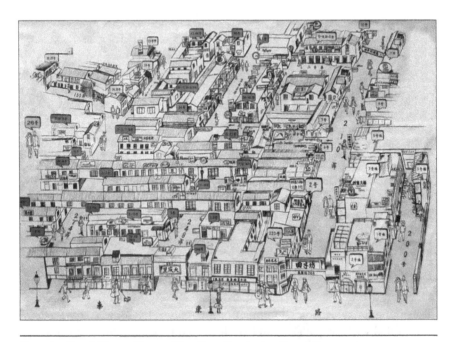

Figure 2 Map of Alleys and Lanes of Tianzifang, Showing Locations of Shops
Source: Map displayed near an entrance to Tianzifang, photographed by Xiaohua Zhong.

workers and other lower-income residents who were displaced by massive redevelopment in central Shanghai. Commercial facilities were lacking in the new apartment houses although a large supermarket, a branch of the French chain Auchan, opened nearby in the 1990s. In response to the demand for local shopping and services, Minxinglu came into existence in an unplanned way and has remained socially and economically viable by serving the local residential neighborhood, as well as visitors and passersby. Most of the inexpensive shops and restaurants are owned by migrants from other parts of China. The one absence is small grocery stores, which disappeared after the Auchan supermarket opened a few blocks away.

These two shopping streets are strikingly different in significant ways. Tianzifang is an interconnected nest of small alleys that one enters through three arched gateways from Taikanglu, while Minxinglu is a linear street that takes up one very long block. Tianzifang has a very strong cultural and artistic orientation and is definitely a "destination" street, while Minxinglu caters to the everyday needs of local residents.

Moreover, Tianzifang is generally high-end or upscale in terms of the taste and pricing of its shops and the wealth of its shoppers, while Minxinglu is low-end or downscale in products and prices. Tianzifang is global in the origin of some of its shops and its ability to attract foreign tourists. Minxinglu however features only locally oriented shops and caters exclusively to local residents. Tianzifang's central location, as opposed to the more peripheral location of Minxinglu, adds to these sharp contrasts.

Despite all these differences, the two shopping streets show a striking similarity: both depend on commercial development from below, a crucial element of Shanghai's urban history.

Commercial Development from Below

The deep historical roots of local shops in Shanghai offer the backdrop to the resilience of Tianzifang and Minxinglu as local ecosystems. In pre-1949 Shanghai, commercial development originated in street-level shops and alleyway houses financed by the meager private capital of small family entrepreneurs. They often had to pull resources together by borrowing from family members and local money shops in order both to start and to sustain their retail businesses.

While many shops lasted, some did not. But their birth and death, as well as their struggle to survive, created the prevalent local narrative of commercial development from below. Not long after 1949, Shanghai's grass-roots private commerce was severely eroded, if not completely eliminated, by the state, which nationalized all private capital and businesses (see Table 1). But history came back alive around 1980, when market reforms began to allow private commercial development.

Table 1 Small Shops in Shanghai, 1949–78 (Selected Years)

Year	Number of Shops	Shops Per 1,000 People
1949	246,000	70.3
1957	129,900	20.5
1962	31,300	4.9
1978	14,900	2.1

Source: Xiong (1999: 197).

In 1978, state-owned and collectively owned enterprises accounted for just about 100 percent of the total retail sales in Shanghai. By 2013, this pattern was reversed, when the combined share of the non-state retail sector rose to 96 percent. In the most meaningful comparison, the state sector's share of retail outlets dropped sharply from 73 percent in 1978 to 4 percent in 2012 as the share of individually owned stores rose from 0.4 to 15 percent (see Figure 3). The difference between these numbers reflects the presence of fully foreign-owned or joint venture chain stores and supermarkets, as well as those with other ownership forms (Shanghai Statistical Bureau 2013). With this dramatic shift, local shops have become privately owned across the board, and have created a thriving city-wide commercial landscape.

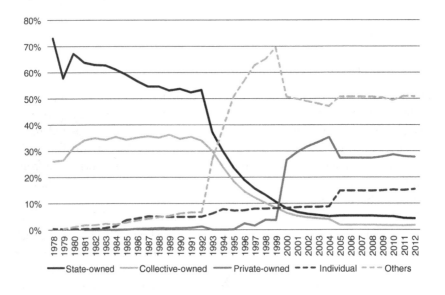

Figure 3 Types of Ownership of Retail Stores, 1978–2012
Source: Shanghai Statistical Bureau (2013).

If the state set commercial development from below in motion, globalization has accelerated and localized it, both directly and indirectly. With incomes rising, Western-style consumption has stimulated the interest and desire of wealthy locals to buy expensive brand-name goods (Sun and Chen 2009). This has led both large transnational retail corporations and small boutiques to set up local branches in Shanghai.

But a part of this global infusion also involves overseas Chinese entrepreneurs or local entrepreneurs who have studied abroad or are, at any rate, globally oriented. Some of them have opened stores with a hybrid approach, catering to both modern western and traditional Chinese tastes. For example, a boutique in Tianzifang sells expensive clothes to a specialized niche of white-collar professional women. Because the design and material of the clothes in the shop reflect the distinctively simple style of ethnic minorities in the mountains of southwestern China, the products have a strong appeal to wealthy female consumers who may be globally oriented but still prefer more traditional Chinese fashions (more on this shop later).

The rapid growth of large-scale shopping malls and supermarket chains has squeezed out most state-owned stores. In fact, malls and supermarkets are the main shopping destinations for the growing middle and upper-middle classes. Yet, popular as they are, supermarkets have not prevented the re-emergence and even expansion of neighborhood shops. If anything, local shops have become more plentiful and more vibrant. They satisfy residents' need for convenient daily shopping, especially low-income residents who do not have cars to drive to supermarkets for weekly shopping.

The retreat of the state in the retail sector, in conjunction with the deep penetration of global commerce, has opened opportunities for local shops and migrants to staff them. The shops selling foreign-brand goods would not be successful without immigrant investors and operators as local agents.

In the specialized case of Tianzifang, overseas Chinese and foreign expats were crucial to its growth as a "creative" shopping zone with a global flavor, at least during its earlier development. Expat business owners there include Australians, French, Japanese, and Taiwanese, recalling the many European business owners in Shanghai during the first half of the twentieth century. While their overseas background may distinguish their shops from those owned by Shanghai businesspeople, they have often tried to adapt their décor to the local environment and culture.

In contrast, domestic migrants from poorer parts of China dominate the shops along Minxinglu. These include a barber shop owned by a young man from neighboring Jiangsu province, a noodles shop owned by a Muslim family from Qinghai province in northwest China, and a metal workshop owned by a couple from Jiangxi province near Shanghai. Again, the prevalence of migrants in local shop ownership and operation brings back a salient feature of neighborhood commerce in pre-1949 Shanghai, except that today's shops are officially sanctioned, albeit lightly regulated, by the local govern-ment.

On both shopping streets, as elsewhere in the city, the spatial foundation of bottom-up commercial development is the conversion of residential units to commercial spaces, usually by the owner of the ground-floor apartment or each of floors of a traditional three-story house. The change to the built environ-ment may be sanctioned and facilitated by local government, after it has been conceived, built, and carried out. But significantly, the local government does not reject it. This unique process of bottom-up commercial development distinguishes Shanghai from most other global cities, while highlighting the common forces of globalization and migration that have shaped all the cities in this book.

Utterly unlike except in their commercial development from below, Tianzifang and Minxinglu are nonetheless connected by the state's policy of large-scale demolition and new construction in the center of the city, with the resulting displacement and relocation of longtime, low-income residents toward outlying districts. But while Tianzifang's commercial development from below was able to prevent demolition and keep some longtime, working-class residents in their homes, many of Minxinglu's shoppers from the local commu-nity are residents of other neighborhoods in downtown Shanghai who have been displaced.

Tianzifang: From Artistic Zone to Commercial District

In contrast to the large-scale demolition and redevelopment that has turned the old, low-rise quarters of Shanghai into tall residential towers and large shop-ping malls, Tianzifang represents a rare exception of small-scale, cumulative, and managed urban regeneration. This process was initially inspired by a visit of a few government officials to SoHo, in New York, which led local actors to adopt a focus on the arts and creative industries. Gradually, however, Tianzifang turned into a special shopping street known for its heavy concen-tration of the ABCs of commercial gentrification: art galleries, boutiques, and cafés, very much like SoHo's evolution. However, unlike in SoHo and other artists' districts, this has not started the kind of residential gentrification gener-ally seen in large American cities like New York (cf. Zukin 2014).

A Low-Key but Auspicious Start

Before becoming Tianzifang, the Taikang Road area was a traditional Shanghai neighborhood comprised almost exclusively of the typical two- or three-story *lilong* houses of Shikumen style, surrounding a small number of one-story workshops and small factories. Arrayed along alleys called *long* (*long* pronounced in Shanghainese and *nong* in Mandarin), the houses resemble

Anglo-American terrace houses or townhouses, distinguished by high brick walls enclosing a small front yard and strong, dark-colored gateways known as Shikumen (or "Stone Warehouse Gates"). Once accounting for more than half of the total housing stock in Shanghai by the 1930s and housing almost 80 percent of Shanghai's population by 1980, Shikumen houses had largely disappeared because of extensive demolition through state-led large-scale urban redevelopment within and beyond the historic center during the 1980s and 1990s, before Tianzifang's birth.

Tianzifang came into existence under a set of fortuitous, though threatening, macro conditions in Shanghai in the 1990s. Market reform coupled with rising production costs led to major industrial downsizing of state-owned enterprises and nearly one million factory workers being let go. With only 7.5 square kilometers but 420,000 permanent residents, Luwan District where Tianzifang was located had a very high residential density, numerous dilapidated traditional houses, and hundreds of old and inefficient factories. This created both pressure and opportunities for shifting from manufacturing to services, utilizing surplus labor, and improving the built environment (Su and Pang 2014).

At the same time, a powerful force behind the eventual strong presence of street commerce at Tianzifang was the regeneration of Shanghai's old central districts. It was motivated by the goals of building Shanghai into a global city, improving its physical image, readjusting the spatial structure of the local economy, and generating more revenues for the municipal and district governments. This led to extensive demolition of Shikumen houses and the relocation of around three million residents from their old neighborhoods. Many small shops that had been embedded in the alleyways disappeared, and large, luxury shopping malls were built.

However, due to the Asian Financial Crisis in 1997–98, large-scale real estate investment came to a temporary halt. Plans for demolition and new construction in central Shanghai, led by the municipal government, were put on hold. This gave the sub-district government an unexpected opportunity to experiment with small-scale urban renewal using the arts as a driver.

Having visited SoHo in lower Manhattan himself, the then Communist Party secretary of the sub-district government attempted to lease the idle neighborhood factories to cultural producers. Mediated by a local businessman who was inspired by a visit to Vancouver (Greenspan 2014), this initiative was ignored, or at least it was not opposed, by the higher-level district government. In fact, there was a tendency on the part of some district governments including that of Luwan to give some autonomy to the sub-district government or the Street Committee in deciding on how to reuse idle factory spaces and redeploying laid-off workers.

Because the main street, Taikang Road, was full of vehicle traffic and would not be so attractive to pedestrians, a decision was made to select an inside lane to create the initial cultural spaces. This would be done without any large-scale planning for redevelopment. In 1999, at the invitation of the local Communist Party secretary, Chen Yifei, a well-recognized Chinese painter, moved his art studio into a simply renovated factory building located at no. 2 on Lane 210, off Taikang Road (see Figure 2).

Chen's arrival drew other Chinese artists to rent the adjacent factory spaces. Their high ceilings and building aesthetics also attracted some foreign artists. In 2001, after visiting Chen Yifei Art Studio, the distinguished artist Huang Yongyu gave a historical literary name to this growing area of artists' studios and cultural firms: Tianzifang, which refers to a piece of land where artists, designers, and scholars gather. Many intellectuals supported this initiative.

Both the name and the cultural reference got a good reception, and in 2005 Tianzifang received post-facto official designation by the city government as a "creative industries cluster" (see www.huangpuqu.sh.cn). In the next few years, with more studios, shops, and shoppers arriving, Tianzifang got more official recognition as a "creative industry park." It became an influential "brand" of its Shanghai district.

Yet the area was already becoming a different kind of shopping street.

Broader Commercial Development

As artists and other cultural entrepreneurs grew their businesses, Tianzifang took off as a vibrant cultural and tourist shopping destination. Its success as a small-scale, market-driven, and largely unplanned development stood in contrast to the dominant mode of government-orchestrated urban redevelopment sweeping across Shanghai. Yet Tianzifang's good fortune drew growing interest in, and pressure for, planned development of high-rise residential real estate from the district government. Clearly, this would threaten demolition of the old Shikumen houses, relocation of local residents, and displacement of the thriving new artists' cluster.

In the meantime, the small number of old factory spaces had all been leased (see Figure 2), which began to limit the expansion of Tianzifang's art scene and cultural industries. Demand for storefronts led local residents who had the good fortune to own ground-floor apartments to seize the opportunity in leasing their residential spaces to commercial tenants. In 2004, their actions propelled the ABCs into the residential alleyways next to the fully occupied factory spaces.

As more artists renovated and settled into new commercial spaces in the old Shikumen houses, they updated and adapted the spaces to fit their artistic tastes and innovative uses. Some added colorful signs and door fittings on the façade of old building structures (see Figure 4), injecting a more animated atmosphere into the narrow alleys (Shinohara 2009).

Once the floodgate of non-residential use was opened, even more apartment owners wanted to lease their spaces to retail businesses. It gave them a voice and

Figure 4 Renovated and Decorated Shikumen Houses in Tianzifang
Source: Photo by Xiangming Chen.

stake in the growth and success of these shops that would allow them to make money from rents. They planned to cash in on Tianzifang's growing reputation as a cultural shopping district and on rapidly rising, market-rate rents, much lower than the subsidized prices they had paid to buy their homes.[1]

Beginning from no leased space in 2004, one-third of the two- or three-story Shikumen houses were renting space to commercial tenants in 2008. Two years later, more than half the houses were renting commercial space, and by 2013, nearly 90 percent were doing this. Even more important, individual floors of these houses, owned by different landlords, were leased to separate businesses (field interviews, June 2014).

As a result of the increase in commercial space, the total number of businesses in Tianzifang rose from zero in the year 2000 to more than 660 by 2013. The number really soared from 2006, after Tianzifang was designated a cultural industries district. Significantly for commercial development from below, more than 400 of the new businesses were located in the Shikumen houses rather than in the factories where Tianzifang began.

Moreover, the non-arts businesses grew faster than those in cultural industries. According to data we collected in the field, 51 percent of all the businesses operating in Tianzifang in 2013 were retail stores of different kinds, 17 percent were restaurants and cafés, and only 29 percent could be categorized as in arts

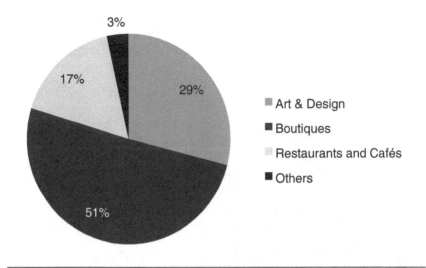

Figure 5 Types of Business in Tianzifang
Source: Walking census, 2013.

and design, with three percent "others" (see Figure 5). In contrast to the more recent commercial conversions, 52 percent of the "first generation" of businesses in the factory zone were in arts and design, 25 percent were restaurants and cafes, and only seven percent were various kind of retail stores, with 16 percent "others."

Tianzifang's extensive commercial landscape bears a strong imprint of globalization and migration. In 2008, 24 percent of all the registered businesses were either completely or partly owned by Australians, French, Japanese, and residents of other countries; 14 percent were Taiwanese owned, and three percent were owned by residents of Hong Kong. Only 21 percent had local Shanghai ownership, while 25 percent were owned by people from the rest of China. The national identities of the remaining 13 percent of owners were unknown (interviews, 2010).

Rising Rents

Given the traditional Shanghai-style architectural environment of Tianzifang, its global connections are undoubtedly structured, on the one hand, through the international business owners and what they sell or serve, and, on the other hand, Shanghai consumers' strong desire for global brands (Sun and Chen 2009). While this lends credence to Tianzifang as a globalized habitus as discussed in the introduction to the book, the shops and restaurants in

Tianzifang vary in the relative weight of their domestic and international roots and ties. Although many shop owners perceive and position their businesses as "global," others see and position them as "local."

Café Dan, a relatively high-end and sophisticated coffee shop set in a tastefully renovated Shikumen house in the heart of Tianzifang, is one of the "global" businesses, yet it is constrained by "local" business factors (see "Shopkeepers' Stories" section on Café Dan, page 74 below). After the initial five-year lease reached an end in 2012, the next lease saw their rent more than doubled. Demand for Tianzifang's unique reputation and location has exerted a huge upward pressure on rents, and no rent controls are imposed by the city government. Moreover, in Café Dan's case, three different landlords own the three small floors that the business occupies, making negotiations difficult, if not chaotic.

Café Dan is not facing dramatically rising rent alone. Our field interviews show that the monthly rent for the ground floor of a typical house in Tianzifang, 24 square meters (258 square feet), rose from RMB500 (less than $100) in 2000 to RMB30,000 ($5,000) in 2012.[2] During our follow-up interviews in 2014, rents for a few similarly sized commercial spaces shot up to as much as RMB50,000 ($8,000) a month, which has forced more turnovers of shops.

The root cause of the rapidly rising rent is commercial development from below, the very factor that has protected local residents from displacement. When Tianzifang's residents were allowed to illegally turn their houses into non-residential uses, it set off an explosion of rent profiteering. With new shops competing to open in Tianzifang, local apartment owners, most of whom no longer live there, kept pushing up the rent, making it increasingly difficult for artists and art-related businesses to survive. Ironically, working-class residents who bought their apartments when enterprise ownership was ended by the state have become absentee landlords, and they are putting Tianzifang's cultural ecosystem in jeopardy.

High rents have a dual impact on the small shops and cafes in Tianzifang with strong global connections. On the one hand, high rents make the entry cost prohibitive for less profitable, locally oriented merchants who might pose competition or bring down the cachet on which higher-status, specialty businesses depend. On the other hand, rapidly rising rents threaten to squeeze out profitable, globally oriented businesses like Café Dan.

This experience is confirmed by the owner of a clothing boutique in Tianzifang called Urban Tribe (see "Shopkeepers' Stories" section on Urban Tribe, page 76 below). When she opened her store in 2007, the same year as Café Dan, she paid RMB1,500 a month. But when the lease expired five years later, her landlord raised the rent fourfold to RMB6,000 a month. Because she does not want to move from a good location, all she can hope for is that the city government will try to impose rent controls.

A Flexible Local State

The clothing boutique owner's hope for relief from rising rent reminds us that the state heavily influences the ecosystem of local shopping streets through its many regulations and policies. While the state has influenced shopping streets in New York and Amsterdam through welfare and tax policies and by enforcing local zoning laws, the presumably stronger Chinese state plays a more differentiated and flexible role.

In Tianzifang, at least at its earlier stage of development, a low-level government official was the prime initiator of change, completely contradicting expected redevelopment plans for the area. By endorsing a "romantic" vision of Tianzifang as a cultural zone and encouraging the lease of factory spaces to artists, the Party head of the sub-district government stood up to the district government, albeit not confrontationally, and won the support from the latter's Party boss, thus preempting the dominant approach to large-scale urban redevelopment through demolition.

The government official also discreetly mobilized a few prominent intellectuals in Shanghai to write and speak out about the need to preserve the historical architecture of the Taikang Road area. This tactic happened to dovetail with the municipal governments use of "significant architectural heritage" as a marketing tool to brand Shanghai's history and attract tourists (Arkaraprasertkul 2013).

Following its initial flexibility in allowing the arts to trigger development, the lowest (sub-district) level of government then tacitly sanctioned and subtly directed the conversion of residential space to commercial use by building owners. This solidified the model of commercial development from below, well known from the city's past, and brought quick economic benefits to the building owners, who became a constituency for further commercial development.

This allowed the relatively weak sub-district government to win a tacit approval from the stronger district government. Remember that it was a sub-district Party official who had originally visited SoHo and played an important role in getting Tianzifang started. This suggests that, in some circumstances, the local state can be so moved by the vision and action of enlightened individual officials that they are able to circumvent higher-order plans and regulations regarding urban redevelopment.

Yet the local government has also been required to respond to the needs of Tianzifang's shop owners that stem from earlier uncoordinated, market-driven growth. Café Dan's owner spoke about the need for new infrastructure and streamlined regulations:

As the rent is going up this much, it is good to have the government more formally involved … We didn't have enough of a power supply and suffered brownouts. The government has taken care of it. The government has also upgraded the old sewer system and the fire protection system. In addition, the government has formalized the procedure for approving operating permits. We are grateful to the government for having cleared a number of hurdles for us to do business here in Tianzifang.

The owner of Urban Tribe echoed this opinion. "Tianzifang used to lack a good and complete service infrastructure because there was no planning at all and every business owner or operator fended for herself. Now the key is to deal with the broader issues better to maintain the original creative spirit that drew many artists and business entrepreneurs to Tianzifang in the first place."

While the local government has stepped back in to bring some order to Tianzifang, it is not significantly altering the shopping street's entrenched and somewhat distorted market logic. Indeed, the state's improvements to the utility infrastructure and permit procedure facilitate rather than moderate market forces. Rising rents have displaced some of the early creative entrepreneurs or at least including a prominent photographer who had started with the painter Chen Yifei, rising rents push them to change their business model. And while some of the initial commercial tenants have left because they cannot afford to pay high rents, they have been replaced by shops selling mass-market merchandise.

These new merchants tend not to appreciate Tianzifang's artistic origin and creative business orientation. In the view of some of the early business owners, the new merchants are only speculators who strive for profits. The owner of a well-established arts store told us:

The situation of copying is getting very bad. Other people sell what you sell and do not even think about it. If you have good sales and work hard to develop new business strategies, other competitors will copy them right away. Replicating and duplicating people's business success has become widespread here. They like to walk around in your shops, finding out what you sell very well, and then they can and will sell that too, often at a lower price. This is a serious problem of intra-industry competition.

As commercial development from below has entered a new and more competitive phase, the number of original and signature shops in Tianzifang have shrunk, accompanied by a reduction in the number of shops owned by foreigners. The crowding-out effect produced by the rising rents diminishes Tianzifang's most appealing attributes. The departure of the creative shops and other specialized businesses will ultimately lead to the disappearance of the consumers who enjoyed Tianzifang's original ambiance and allure.

Shopkeepers' Stories

Café Dan, Tianzifang: Global City, Rising Rent

Café Dan opened in 2007 and quickly become widely known for its exquisite specially brewed coffee and excellent Japanese food. The owners are a married couple, one Chinese and the other Japanese, both with experience working in transnational environments. The wife is a native Shanghainese who had previously worked at a famous local hotel and then went to Japan in 2002. She married a Japanese man who had worked as an engineer for Texas Instruments.

The idea to open a cafe in Tianzifang came to Mrs. Dan when her friends brought her there on a return trip to Shanghai in 2007. She was favorably impressed by the preserved traditional Shikumen architecture and the already strong international atmosphere. At that time, there were relatively few good specialty coffee shops in Shanghai and hardly any in Tianzifang. Given her husband's love for high quality home-brewed coffee, Mrs. Dan knew about some small coffee shops in Japan serving family-brewed specialty coffee, and thought they could build on her husband's love of coffee to open a business like that. The couple felt that Tianzifang was the best place to open a special kind of café. "Tianzifang is a cultural place, attracting many international customers. The place is ideally suited to open a nice café."

Since the beginning, Café Dan has positioned itself between the global and local. Its coffee not only comes from Africa, but also from China's southwestern province of Yunnan. It not only accounts for almost all coffee production in China, but also grows a special indigenous kind of small coffee beans that is popular in Japan. The couple's initial investment was not as large as it might have been if they had opened a café in Japan, but it wasn't as small as it would have been in other areas of Shanghai, because of Tianzifang's higher rents.

While Café Dan's owners love the large number of international customers, they told us that about 70 percent of their customers are Chinese, and predominantly Shanghainese, including a number of famous actors and high-level government officials. This shows how Tianzifang provides a globalized habitus to local customers.

Although Café Dan caters to globally oriented, high-income local consumers who can afford a $4 cup of coffee, the owners say their expenses are high.

We don't know how well our direct competitors do, but our costs are higher because we use really good imported coffee beans and quality ingredients for our Japanese dishes. While some restaurants may keep using the same oil for cooking, we use fresh olive oil. By serving both excellent coffee and authentic Japanese food, we attract a variety of customers including older people and children. Our set or packaged lunch costs about RMB100 [$16], which is not cheap even in Shanghai.

The owners have been trying to expand their space and offerings to capture more profit:

> We have recently expanded the size of operation by turning the first floor, which was used for storing coffee beans, into an additional space for customers. We have also begun selling liquor on the premises.

But the owners are able to economize through their hiring practices. They told us they typically have around ten workers who are migrants from poor interior provinces like Sichuan and Henan. These workers are paid RMB12–15 ($2–2.50) per hour, which averages to about RMB3,000 ($500) a month.[3] Since the building is small in floor area, the owners provide neither living space nor free meals but instead give RMB15 ($2.50) per day to each worker as a food subsidy. The workers also get a small bonus every month based on their performance and total sales. To keep payroll costs down, both owners put in a lot of time themselves working up front and in the kitchen.

Despite its trendy and global features, Café Dan is constrained by local rents. Like the high rents charged by building owners in gentrifying neighborhoods of New York and Amsterdam, Café Dan's rent is high by Shanghai standards and rising fast. Not only do building owners in Tianzifang take full advantage of the unique location, the ownership structure is somewhat chaotic.

To get a five-year lease in 2007, the owners of Café Dan had to negotiate with three different landlords who separately owned the three floors of the building,[4] while the owners of the first and second floors argued with each other. When we interviewed Café Dan's owners in 2012, they had just renewed their lease for another five years but had to accept a doubling of the rent. Although this increase may be common in New York and other global cities, it is still a new situation in Shanghai.

The wife lamented that this increase must be irrational for a building owner:

> I don't understand why the building owners would want to kill the chicken that lays the golden egg. If successful shop owners are unwilling to pay the much higher rent and leave, will the building owners be able to find good renters like us? You may demand an astronomically high rent and be lucky to get someone to 'bite,' as many small businesses desperately want to set up shops in Tianzifang. But can you guarantee to have a reliable and profitable business like us who can stay beyond just one or two years?

As high as the rent is, however,

> It is not too bad for us as we deal with primary landlords, not those who sublease other property owners' spaces to shop owners. It is ok if you raise the rent after five years because we are doing well and prices are going up everywhere.

In 2014, the owners of Café Dan acquired another piece of the first-floor space, which allowed them to add a closed-in second-floor balcony to expand their business a little further. While this has added new costs to the already much higher rent, it is a clear sign that the owners are determined to stick out and make it in Tianzifang for the long haul. They have also ruled out any plan to open up another café elsewhere to create another revenue stream. It reflects their belief that they will be able to sustain a viable business based on their strengths.

Urban Tribe, Tianzifang: Rising Rent, Unique Location

The owner opened this women's clothing boutique in 2007 after working at a foreign-owned company and as a freelancer in Shanghai for almost 20 years. Her business partner designs clothes with a rugged look that reflects the colors and styles worn by minority groups living in the mountains of Yunnan province in southwestern China. The owner markets the traditionally styled and finely made clothes mainly to international customers, including some from Taiwan and Macau. Although the business was profitable, by 2012, she was very concerned about the rising rent.

> When I started in 2007, I paid RMB10,000 [$1,600] a month for my one-level [first-floor] store space with a five-year lease. Now I have to pay RMB40,000 [about $6,500], a four-fold increase. If I pay all of it upfront, it can kill my cash flow.
>
> At this exorbitant rent, I can move to Xintiandi [the most famous luxury shopping area in Shanghai, which is not far from Tianzifang], but I won't. I chose Tianzifang originally because the rent here was reasonable and the place was known and conducive to developing creative ideas and products, although it was not very well planned. Since I have been making money … I will try to stay in Tianzifang and see if it will continue.
>
> Another reason to stay is that we were the first to set up shop on Lane 248 and have become its landmark. This is why customers keep coming back to buy our products and the media have promoted us where we are.
>
> Since the rent has gone up so much, I may have to adjust my strategy by coming up with a few less expensive dresses. I will try very hard to stay in Tianzifang. I also hope the local government will do something to offset the very high rent.

Minxinglu: An Ordinary Shopping Street

In contrast to Tianzifang's creative businesses, Minxinglu is an ordinary shopping street that provides for the everyday needs of local residents. There's no cocktail bar or special café, but there are fruit and vegetable shops, inexpensive shoe stores, and a casual mahjong parlor.

Urban Context and Origin

Minxinglu is located in the Zhongyuan area of Yangpu district in northeast Shanghai, quite some distance away from Tianzifang. As early as the 1950s and 1960s, the Zhongyuan area was built up with new apartment houses for workers, which reflected the post-1949 expansion of Shanghai as a major manufacturing center dominated by state-owned enterprises (SOEs). As usual in China in those days, much of the housing was built by SOEs as low-rent apartments for their employees.

The 1980s brought both continuity and change. While there was continued large-scale residential construction in the area, including the SOE welfare housing, apartments were also built for the initial wave of residents relocated from the demolitions of central Shanghai. This process accelerated in the 1990s when more and more people moved to Zhongyuan after losing their old houses to demolition and being displaced.

Today the area is the most populous sub-district in Shanghai with more than 300,000 residents in eight square kilometers. Given the history of Zhongyuan, we see Minxinglu as an indirect product of the larger forces of globalization, migration, and state-directed urban redevelopment that also shaped—and threatened—Tianzifang.

As a relatively new and rapidly expanding residential area in the 1990s, Zhongyuan had a severe shortage of commercial facilities relative to its large population, which is typical of most newly developed residential districts in Shanghai. Unlike Tianzifang, which caters to more cultured and wealthy shoppers and tourists, stores in Zhongyuan have to meet the needs of mostly retired workers and government employees, other lower-income people, or just passers-by through the neighborhood.

As in Tianzifang, however, the small number of shops on Minxinglu in the 1990s began through commercial development from below. The local government allowed owners of ground-floor apartments to convert, or build out, and later lease, parts of their space for commercial use. At first, almost all of the shop owners in Minxinglu were local residents with Shanghai *hukou* (household registration), most of whom were the apartment owners who built the storefronts. Many of them opened convenience stores of various kinds (Figure 6).

It was not long before these small shops would face overwhelming competition from an Auchan Supermarket, which was the first comprehensive supermarket opened in China by France's Auchan Group. With registered capital of $18 million, Auchan occupied 21,000 square meters, truly an impressive marketplace that may be a thousand times larger than a shop in Minxinglu. After opening in 1999, Auchan expanded in 2003 to 24,000 square meters and added 300 parking spaces on a new second level and a food plaza. With low prices and service advantages, Auchan forced many local grocery and

Figure 6 Convenience Store in Minxinglu
Source: Photo by Xiaohua Zhong.

convenience stores to close or change owners. For the most part, the new shop owners are migrants from rural areas and interior cities in China.

Aside from several greengrocers and a fish store, there are no food stores left on Minxinglu that would compete with the huge Auchan supermarket. Nonetheless, there is a broad diversity of stores typical of most shopping streets in residential neighborhoods.

Diversity of Shops

Unlike the lanes of Tianzifang, Minxinglu is a straight road that stretches along the southern part of Zhongyuan district. Looking at a part of the road, the north side of the 300-meter-long commercial block between Baotou Road and

Zhongyuan Road, we count 99 shops. Of these, 25 percent were built as commercial properties or properties that belong to *danwei* or government-owned work units, while most of the rest are unauthorized conversions of residential units into shops, which were made, as in Tianzifang, by the residents. Aside from a small number of residents who operate the shops themselves, the overwhelming majority (95 percent) of the shop owners are migrants from outside Shanghai.

While Tianzifang's shops are set on both sides of narrow *lilongs* that form an irregular grid or maze of preserved traditional houses, set off by gateways from the surrounding roads, the shops on Minxinglu line up in a straight row facing the sidewalk and a fairly wide street with busy traffic. Walking along Minxinglu, one is struck by the diversity of shops that carry colorful signs to advertise their businesses. But there is little doubt that they sell mostly inexpensive goods catering to the everyday needs of lower-income people.

Of the 99 businesses on Minxinglu, most are restaurants (26), clothing and shoe stores (23), and food stores (12), with the rest providing building materials (6), paper and pencils (6), mobile phones and hardware (5), manicures (4), printing and CD burning (3), haircuts (3), foot massages (2), fruit and vegetables (2), and repairs of household appliances (2). There are also two small motels, and one each of a tailor shop, photo studio, chess and mahjong room, and pornography shop (see Figure 7).

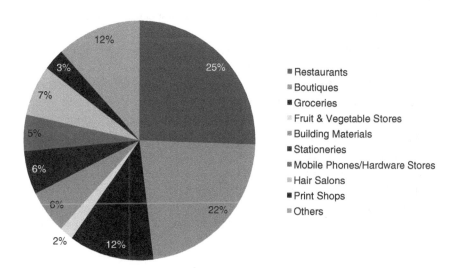

Figure 7 Types of Shops on Minxinglu
Source: Walking census, 2013.

Lives Completely Altered

Its mundane status aside, Minxinglu, through the life experiences of its shop owners, tells a story of the dramatic transformation of Shanghai since 1990. We would never learn this story if we only visited Tianzifang.

We will use two shopkeepers' stories to illustrate how the lives of ordinary residents have been completely altered by urban redevelopment, beginning with the story of the convenience store owner shown in Figure 6.

Shopkeepers' Stories

Minjia Convenience Store, Minxinglu: Status, Service, and the State

Born in 1950 in Shanghai and only a high school graduate, Mr. D is both a local resident and the proprietor of *Minjia* (People's Good) convenience store. He belongs to the Cultural Revolution generation, whose members were sent to labor in the countryside and in factories, and thus missed a chance to get a college education. Although two of his three sisters were sent to farms in Heilongjiang province in northeastern China, Mr. D was assigned to work at the state-owned Shanghai Fishing Boat Factory. He married a woman who worked at the same factory in the late 1980s and was given an apartment on the ground floor, facing Minxinglu, as a welfare benefit from his factory.

In 1990, when his employer was not doing well financially, Mr. D was furloughed with basic pay but no bonus. The lost income prompted him to turn the courtyard of his apartment into a storefront, put a roof over it, and open a grocery store.

Mr. D talked about local conditions when he first came to Minxinglu:

> When I first moved to Zhongyuan in the late 1980s, it was just being built up. There was not a single new building across the street. What was there was a steel and iron warehouse, and Minxinglu was very narrow.... When I opened my store in 1993, there was only one other store owned by a displaced resident from central Shanghai. The buildings behind my store were not even fully occupied.
>
> Early on, I sold cigarettes, stationery, and other sundry goods that filled the shelves. Now I only sell cigarettes, and nobody bothers to buy the soft drinks, so we consume them when my son comes here. While some neighbors who moved into these apartment buildings with me have left, I have stuck around and seen a lot of changes around this community.

Since local shops in other cities operate with licenses and varied support from the local government, we fully expected Mr. D to have had the same experience with the powerful local government in Shanghai. But the account Mr. D gave us is more complex and nuanced, and also differs from the commercial development from below in Tianzifang.

> In planning to open my store, I had to go to the Yangpu District government to apply for a license; otherwise I would not be able to have food on the table. I was lucky then because two or three months after I got my license, the local government stopped issuing licenses.
>
> Since I was converting my residential space to commercial use, I also had to go to the district housing management bureau to get approval. They would not approve at first, but I began to open up the walls anyway.
>
> They came to stop me and told me that if I could obtain an operating license [for the business], they would approve [the building conversion]. But the district industrial and commercial management bureau told me that they would issue the license [only] if I could be approved by the housing management bureau.
>
> In the end, I dragged someone from the housing management bureau to the other bureau. They had to discuss my case face-to-face and agreed to give me the operating license.

Mr. D's difficulty in getting his operating license may suggest that the local state takes a rigid and strong role in regulating privately owned shops. At least, it suggests that local government agencies act like a typical bureaucracy. But this is not the case: most of the shops on Minxinglu that opened after his have been operating without a license. The reason, according to Mr. D, is that the local government agencies do not want to monitor these shops, and not issuing licenses is their way to dodge responsibilities for enforcement.

> They [the government] did not care about the shops that have opened since my shop opened. Most of the owners were laid off SOE workers who had to make a living. There was no specialized municipal affairs agency to regulate the shops because they were inside the *danwei*-owned housing. But local officials tried hard to collect management fees even from owners like me who were licensed.
>
> Once there was a water leak inside my shop. I went to the housing repair bureau to ask them to fix the problem, but they refused to do so. So I challenged them on what they had collected management fees for, and we had a big quarrel.
>
> When the Industrial and Commercial Management Bureau renewed my license, they tried to find fault with this and that. When I spoke with them in a strong voice, they treated me as a bad person. Later the municipal affairs agency required me to buy plastic pails for storing waste, but I think they should pay for it.

In starting his shop and keeping it going, Mr. D has both benefited from his former status as a SOE employee and suffered from the state's bureaucratic tendencies. If he had not been laid off by his factory, he would not have been able to open his own store. If he had not got a big discount on the price of his apartment because of his employment service, he could not have "bought" an apartment and would have had to pay rent for the store. But if the local government had made it easier, he might have done better with his shop financially or tried to make it more attractive.

Ultimately, Mr. D has not been able to compete with the Auchan supermarket. And he feels bad about it:

> Shopping at supermarkets is a faddish trend that Chinese follow. I told my wife not go there. The pig's feet sold there do not even taste right.
>
> I have been doing less and less business ever since Auchan opened. Look at today so far, from 6 a.m. when I opened the door until now, 2 p.m., I have only sold RMB70–80 [$12–14] worth of cigarettes. The buyers are basically old customers who work across the street or live nearby.
>
> People go to Auchan to buy drinks and liquor in large quantities at a much lower unit price. Even I would go there to buy a small amount of bottled drinks to resell [in my store]. If nobody buys them, I'll enjoy them myself.
>
> There is also a lot of pressure on the other shops. The commercial space next door, for example, has turned over a number of times. If I didn't own my own store space and I had to pay rent, I would already have gone under.
>
> I could easily rent my shop space for RMB5,000–6,000 [$1,000] a month, but I can't do it because of my license. Renting it out could get me into trouble with the industrial and commercial management bureau. If I didn't have a license, I would have already rented it out. Honestly, I am now content with just selling some cigarettes and making a little money.

FDLC Hair Salon: Migrant Business Owner, Local Customers

The second shopkeeper's life story is different, yet his experience on Minxinglu complements Mr. D's in revealing both harmony and tension in the relationship between the shopkeepers and the local state.

Mr. C is in his early 30s and already owns a hair salon named *Feiduanliuchang* (FDLC, whose four Chinese characters are roughly translated as "fly, short, keep, long"). A migrant, he was born and raised in the city of Yangzhou in Jiangsu province bordering Shanghai.

Yangzhou has a reputation for producing and sending out good barbers to larger cities. In fact, upon graduating from high school in 2000, Mr. C migrated to Shanghai to work at a hair salon owned by a relative. Starting out washing customers' hair, he quickly learned to cut and perm hair as an apprentice. In 2005, he landed the position of manager at another salon. In 2006, he opened his own salon, and in 2011, Mr. C opened FDLC on Minxinglu.

Mr. C says he could get "as many as 50–60 customers a day":

> Between 30 and 40 percent of them are residents living around here. The rest are from other parts of Shanghai, including roughly 10 percent from Pudong [the rapidly growing district of Shanghai east of the Huangpu River], some passersby, a few who come upon recommendations, and a number of regular customers from as far as Hangzhou [a 50-minute ride from Shanghai by high-speed train].

Rising from an apprentice to a boss over a decade, and owning his own hair salon in Shanghai, Mr. C has achieved a high degree of upward mobility for a migrant. But it has not been an easy path. The three-year lease he signed with a private apartment owner on Minxinglu in 2011 costs him a monthly rent of RMB5,000 ($850). He recently renovated and upgraded his shop by investing the money he had earned. Moreover, he employs seven young barbers and hairdressers mostly from his hometown, following a common practice at migrant-owned barbershops in Shanghai.

> I make money every month, but it varies. It costs RMB30 [$5] for a cut and shampoo, including the option of cleaning earholes. When we first started, we would be willing to give a free haircut to the elderly residents around here or accept RMB2 [30 cents] for a haircut. Now we only charge RMB5 [85 cents] and continue to offer very good service to elderly customers. We prefer to visit nursing homes to give haircuts on site. Overall, our profit margin is relatively thin because we use good-brand and high-quality hair products and I pay for training my employees.

Mr. C and his employees show how migrants in Shanghai fill a niche in both the labor market and service economy. Of the almost 25 million people living in the city, about 10 million are migrants without local *hukou*. According to the 2012 Shanghai Statistical Yearbook, of the 1.32 million residents in Yangpu district, 280,000 (28 percent) come from other cities and rural areas.

Since Zhongyuan sub-district is a recently developed and densely settled residential area, there is a huge demand there for all kinds of personal services. Moreover, the commercial spaces created by the working-class owners of ground-floor apartments offer opportunities with low entry costs for migrants to start small shops and service businesses. The ecosystem of Minxinglu as a local shopping street demonstrates how all of these factors come together in a different type of commercial development from below: there are building owners who construct storefronts even without legal authorization, business owners who migrate from outside Shanghai to provide everyday goods and services, and Shanghainese customers who have been displaced from the central districts of the city.

Despite the daily contributions of migrant entrepreneurs like Mr. C to local urban life, they have a hard time fully integrating themselves and their families into permanent residents' status in Shanghai. They face a variety of barriers associated with the absence of the local *hukou* but also go beyond it.

> First of all, for my daughter to enroll in a public primary school in Shanghai, we have to have an official residency permit. Since we are renting a street-level commercial space owned by someone else and an apartment to live in elsewhere in the city, we are not eligible. Although we have worked here for many years, we cannot afford to buy a decent apartment that easily costs over RMB1,000,000 [more than $150,000]. The Shanghai government is putting a high priority on recruiting outsiders with a high level of education,[5] and we are not happy with it. The city should attract people at the high, middle, and low levels of human capital and skills. We have experienced the strong anti-migrant attitude of local police and neighborhood watch guards. It makes a big difference whether you speak Shanghai dialect or not. When you run into a problem, it is easy for a native resident to deal with the police but much harder for migrants like us.

Mr. C went on to recount two incidents as examples of what he perceived as discrimination against him. He talked about one incident this way:

> About six years ago, a group of hooligans arrived at a previous barbershop I owned in a van without a municipal license and started vandalizing the space. They did it for the landlord who had re-leased the commercial space to me from the original owner, due to a disagreement between the landlord and me. I immediately reported it to the police, but they did not get there until after those hooligans had left. They took notes on what I said and told me to wait for a week before letting me go.
>   I heard nothing from them after more than ten days and called them about it. I expressed my grave concern about the existence of mafia-like people who dared to destroy my shop during daylight. In the end, they did nothing, and I had to pay for the damage because the second landlord had paid for the interior renovation.

Migrants' difficulty in social acceptance and integration varies by where they come from and how they are perceived by local Shanghainese. The Uighur owners of a halal noodle restaurant on Minxinglu, a married couple with a new baby, have had experiences like Mr. C's.

The husband and wife are Muslims of the Hui minority group who came to Shanghai from Qinghai province, in northwestern China, in 2009. They opened Authentic Lanzhou Beef Stretched Noodles Shop on Minxinglu in 2010. They also own two other restaurants near Minxinglu, one of them located just a block away. The couple hires seven employees who are either extended family members or from the same place of origin, and all the employees live collectively in a residential space behind the restaurant, which is common among many migrant-owned and leased shops and small businesses. They both work and live on the street.

The couple said they feed an average of 70–80 customers every day:

About 60 percent of them are students from the two schools across the street. They come to eat because the stretched noodles are an inexpensive fast food to them, not because they are aware of us as a national minority.

We sell a bowl of noodles for RMB10–15 [around $2], and make about RMB300–400 [$50–65] a day and about RMB10,000 [$1,600] a month in net profit, but we have to pay RMB7,000 [$1,150] per month for renting this space.

We do not interact with the other shop owners on the street; neither do we have any Shanghainese friends. Because we have a different religious belief, we socialize only with our own. We do not see ourselves ever being accepted as members of the local community. We do not have access to public education so we have sent our older children back to Qinghai for schooling.

A Tale of Two Streets in a Changing Shanghai

On the surface, Tianzifang and Minxinglu could not differ more. Tianzifang is located in an old neighborhood of historically preserved and renovated Shikumen houses in central Shanghai, while Minxinglu runs alongside a large block of 1980s-era low-rise apartment buildings in a massive residential district toward the outskirts of Shanghai. Tianzifang is a largely exclusive district of art studios, boutique shops, high-end cafés, and other "creative" businesses. Minxinglu is filled with convenience stores, cheap eateries, service outlets, construction materials shops, and other small businesses. Tianzifang has cultural sophistication, an international reputation, a special allure, and relatively high prices. Minxinglu meets the daily needs of the local community by providing low-cost but convenient, necessary goods and services. In simple contrasting terms, Tianzifang is special, glamorous, and upscale, while Minxinglu is common, mundane, and downscale. They represent two opposites on the spectrum of local shopping streets in a rapidly changing Shanghai.

The sharp differences between Tianzifang and Minxinglu begin to blend if we look at the larger context in which globalization, migration, and the local state shape the ecosystem of the local shopping street. Without globalization and the state-driven project to build Shanghai up as a global city, there would have been no opportunity for Tianzifang to be conceived as "Shanghai's SoHo" and to develop the businesses that confirm that label. Without globalization and the state's opening to foreign direct investment, there would have been no international and overseas investors and entrepreneurs to help launch Tianzifang and give it a global feel.

Domestic human agency through an enlightened local government official and a few prominent artists played an important role in Tianzifang's development. But when the local state tried to impose some order on the area's early spontaneous growth, it really facilitated market forces (Wang 2011). The analytic link between Tianzifang and Minxinglu comes from the state's plan to redevelop central Shanghai and re-settle displaced residents in the new residential district of Zhongyuan. In other words, without Tianzifang, there might not have been Minxinglu.

Large-scale urban regeneration aimed at turning downtown Shanghai into the modern core of a global city spared the architectural heritage of Tianzifang given its appeal to international tourists. But it pushed a large displaced population to new residential districts like Zhongyuan. The new districts' shortage of convenient shopping and service facilities created niches for recent migrants to make a living by opening small shops catering to the everyday needs of local residents. Given the rich history of privately owned shops in Shanghai and the need for jobs for millions of migrants today, the Minxinglu kind of shopping streets would emerge in many parts of the city sooner or later with or without Tianzifang as a special shopping destination downtown.

Tianzifang and Minxinglu indicate that the most salient features of the ecosystem of local shopping streets in Shanghai are the interactions between a flexible and limited local state, the historical legacy of neighborhood shops, and the power of new commercial development from below. Even as the municipal government has driven the rapid build-up of Shanghai, it has sanctioned and supported, albeit belatedly, the cultural and architectural preservation of old houses. This has created a favorable opening and impetus for Tianzifang to play off the theme of Shanghai nostalgia as a new way of remaking the old built environment for shopping, entertainment, and creativity (Ren 2011; White 2013).

While this reflects the active and interventionist side of the local state, officials have been reluctant to preempt or suppress market forces. Local government has not responded to business owners' pleas to thwart the aggressive copying of business products and strategies, dramatic escalation of rents, and entry of standardized, mass-market stores, all of which threaten Tianzifang's original cultural orientation and creativity. Despite the call for the local government to re-energize the creative industry in Tianzifang at a special forum in December 2014, there were hardly any concrete suggestions for what and how this could be done.

If the local state has tried to become more involved in Tianzifang over time, it has not been much engaged with Minxinglu. Since issuing a few operating licenses to laid-off workers like Mr. D in the 1990s, the local government has allowed the shops, almost all of which are owned by migrants, to operate without them. While this has created the appearance and practice of illegality and informality, it is intended by the local state as a strategy for keeping migrants gainfully employed. This hands-off approach has made relatively easy for migrants to open small shops and service establishments.

The local state has recently retreated further by stopping tax collection. Instead it has begun to provide more financial support to migrant-operated shops through municipal budgetary allocation for installation of signage and rain covers over the open store fronts. Yet where the local government really needs to step up is to institute thorough *hukou* and welfare reforms. This is an important factor in strengthening social integration for migrants.

Regardless of what the local state has done and why, what matters most is the interaction of store owners, building owners, and shoppers within the ecosystem of the local shopping street. In Shanghai, these actors have not only kept the spirit of traditional neighborhood shops alive, they have (re)created a more extensive and diverse mode of commercial development from below. Once it is created, this commercial development from below is sustained by the resilient ecosystem of local shops.

The city's ecosystem of local shops has re-emerged from the vibrant pre-1949 history and a dormant socialist period. It has thrived with and against the homogenizing threat of globalization and the driving force of a powerful state. In many ways, the ecosystem of local shops in Shanghai works like those in the other global cities in this book. It is nurtured by hardworking and entrepreneurial (im)migrants to become an integral part of the everyday landscape of urban life.

If there is one distinctive feature setting Shanghai apart from the other cities, it is how local shops have navigated within and through limits set by the state. Local shops benefited from state-initiated market reforms that privatized housing and allowed individual apartment owners to rent part of their living space to retail entrepreneurs. Between the landlords and the business owners, they have recreated a large and diverse commercial space long vacated by the state. The long-run strength of local shops in Shanghai depends on this commercial development from below. Yet it contrasts with state control in Amsterdam.

Acknowledgements

This chapter is based on interviews and ethnographic observations, and a walking census of stores on both streets carried out from 2011 to 2014 by Hai Yu, Xiaohua Zhong, and Xiangming Chen. We express our gratitude and appreciation to all the business owners and managers who generously spent so much time with us. Partial financial support from the Paul E. Raether Distinguished

Professorship Fund at the Center for Urban and Global Studies at Trinity College and the School of Social Development and Public Policy at Fudan University for a workshop in Shanghai helped us to complete the chapter.

Notes

1   The urban housing sector in China through the late 1970s had been owned and maintained almost completely by the government, which viewed housing as non-production and thus a low priority. It resulted in limited housing investment, severe housing shortages, and inferior upkeep. Housing reform in China started as early as 1980 with an initial experimentation with housing sales in selected cities. The second phase of the housing reform in the 1980s involved the government investing more in housing construction, raising the very low rent, and pushing for more privately financed housing development and sales. In 1994, the state began to promote a more unified "commercialized" housing stock, mainly composed of owner-occupied housing. The radical housing reform in 1998 aimed to privatize the urban housing sector within a few years. One important outcome was that almost all the government-owned and subsidized housing units were "sold" to their renters at heavily discounted prices. Those who "bought" them including the residents in Tianzifang could now lease them out to merchants as owners.

2   The ground (first) floor rents for more than the second floor, which has a higher rent than the third floor. Given the very narrow and steep stairs in a typical old Shanghai-style (Shikumen) building, it is easiest to access the first floor, more difficult to get to the second floor, and most inconvenient to climb to the third floor. Although most customers prefer to sit on the first floor of a restaurant or café in Tianzifang, a few may like to climb to the second or third floor to enjoy the nostalgic feeling of being in a tight space of a traditional house.

3   While RMB3,000 ($500) a month is a decent wage, roughly the starting salary for a college graduate these days, it is not enough for living comfortably in Shanghai where the very high housing cost (monthly rent averaging as high as RMB3,000) forces most migrant workers and other young low-wage earners to share less expensive apartment units with others.

4   This crowded ownership dates back to the 1970s when the severe housing shortage in Shanghai forced the residents of these units to sub-divide them to house more families some of whom became de facto owners later and then became landlords who leased their floors to merchants in Tianzifang in the 2000s.

5   Even as the *hukou* (household registration) system has been loosened over time, a few top cities such as Guangdong and Shanghai have used more restrictive *hukou* reforms that use a point system to screen and determine migrants' eligibility for acquiring local *hukou* and receiving social services (Zhang 2012). Despite the even more sweeping *hukou* reform introduced in 2014 that requires all small and medium cities to offer *hukou* to all migrants, the few largest and most popular destination cities for migrants, especially Beijing and Shanghai, continue to set more stringent local policies. This will continue to sustain the hurdles for people like Mr. C to become fully integrated into the city.

References

Arkaraprasertkul, Non. 2013. "Traditionalism as a Way of Life: The Sense of Home in a Shanghai Alleyway." *Harvard Asia Quarterly* 15(3–4): 15–25.

Greenspace, Anna. 2014. *Shanghai Future: Modernity Remade*. New York: Oxford University Press.

Lu, Hanchao. 1995. "Away from Nanking Road: Small Stores and Neighborhood Life in Modern Shanghai." *Journal of Asian Studies* 54(1): 93–123.

Ren, Xuefei. 2011. *Building Globalization: Transnational Architecture Production in Urban China*. Chicago IL: University of Chicago Press.

Shanghai Statistical Bureau. 2013. *Shanghai Statistical Yearbook 2013*. Beijing: China Statistics Press.

Shinohara, Hiroyuki. 2009. "Mutation of Tianzifang, Taikang Road, Shanghai." Paper presented at the 4th International Conference "The New Urban Question – Urbanism beyond Neo-Liberalism" of the International Forum on Urbanism (IFoU), Amsterdam/Delft.

Su, Bingong and Xiao Pang, eds. 2014. *A Close Look at Tianzifang* [in Chinese]. Shanghai: Wenhui Press.

Sun, Jiaming and Xiangming Chen. 2009. "Fast Foods and Brand Clothes in Shanghai: How and Why Do Locals Consume Globally?" In *Shanghai Rising: State Power and Local Transformations in a Global Megacity*, edited by Xiangming Chen, pp. 215–35. Minneapolis MN: University of Minnesota Press.

Wang, Stephen Wei-Hsin. 2011. "Commercial Gentrification and Entrepreneurial Governance in Shanghai: A Case Study of Taikang Road Creative Cluster." *Urban Policy and Research* 29(4): 363–80.

White, Matthew. 2013. *The Tiandi Model: Paradigm for Placemaking in China*. Self-printed booklet by an architecture student as final project for a class titled "Urban Development in China," Yale University, Spring.

Xiong, Yuezhi, ed. 1999. *The General History of Shanghai, Volume 12: Contemporary Economy*. Shanghai: People's Press of Shanghai.

Zhang, Li. 2012. "Economic Migration and Urban Citizenship in China: The Role of Points Systems." *Population and Development Review* 38(3): 503–33.

Zukin, Sharon. 2014. *Loft Living: Culture and Capital in Urban Change*, 3rd editon. New Brunswick NJ: Rutgers University Press.

CHAPTER 4

From Greengrocers to Cafés

Producing Social Diversity in Amsterdam

IRIS HAGEMANS, ANKE HENDRIKS, JAN RATH,
AND SHARON ZUKIN

Amsterdam has long been a global city, marking its "golden age" as far back as the seventeenth century, when the Dutch East Indies and West Indies Companies dominated world trade. Since the 1970s, however, a new era of globalization has reshaped the city in two important ways, and created an ongoing debate about gentrification and social and ethnic diversity.

First, Amsterdam is a major tourist destination, attracting year-round flows of overseas visitors to its canals, beautifully restored historic houses, and rich sites of cultural memory like the Rijksmuseum, Anne Frank House, and Rembrandt's living quarters, as well as to the tawdry (though also very popular) red light district.

Second, the city draws large numbers of transnational migrants who live and work there, and who now make up about 30 percent of the population. The three largest "ethnic minorities" come from Turkey, Morocco, and the former Dutch colony of Surinam, on the northeastern coast of South America.[1]

While the city has become more ethnically diverse in recent years, expanding beyond its historic base of a mostly European population, cultural differences and economic inequalities create tensions around the national policy of social integration. Non-western migrants, especially Moroccans and Turks, have lower levels of education and are twice as likely to be unemployed as native-born Dutch. That many migrants are Muslims, in contrast to the secular, though predominantly Christian, Dutch majority often arouses distrust on both sides. Moreover, public policy in the Netherlands promotes gender equality, open expression of sexual preferences, and tolerance of cultural nonconformity, all of which can be felt as threatening by newcomers from more traditional societies.

Officially, the Dutch state opposes the "ghettoization" of ethnic minorities and all the social problems that word implies. But since the 1980s, Amsterdam has increasingly been divided between a gentrified center, including the prominent Canal Belt, where relatively few migrants live, and large concentrations of ethnic minorities with recent immigrant origins who settle in the Oost (East) and West districts of the city. Separation by income and ethnicity is accentuated by increasing rates of home ownership, with rental apartments in subsidized "social" housing being converted to owner-occupancy, rental tenants forced to move elsewhere, and apartments being purchased by somewhat more affluent, native Dutch (Boterman and van Gent 2014).

Despite a burst of state-planned urban renewal in the 1960s and 1970s, which demolished old houses near the center to build the city's first subway line, and added a new corporate office district and housing in the South, Amsterdam has for the most part kept its traditional streetscape. Four- and five-story red-brick townhouses, some dating back to the seventeenth century, line the narrow streets, and Amsterdammers ride their bicycles everywhere. Small cafés and shops fill the narrow "footprint" on the ground floor of many houses.

Yet since the 1960s, as almost everywhere else in the world, the expansion of big retailers has dramatically reduced the number of traditional small groceries, butcher shops, and hardware stores. Foreign retail chains like H&M have spread throughout the center. Supermarkets and other retail chains specializing in personal care products and inexpensive housewares are found in every part of the city.

Most retail businesses are probably owned by native-born, "white" Dutch people. However, many transnational migrants, who initially came to Amsterdam during the 1970s to work in manufacturing industries, opened small shops and restaurants when the factories closed and their jobs shifted elsewhere. As in New York, Berlin, and Toronto, they tended to locate in working class neighborhoods where rents were low and they catered to transnational migrants like themselves.

Today, Amsterdam's retail landscape shows the impact of two contrasting paths of globalization. The first, cresting on a wave of high-status restaurants and boutiques, is populated mainly by "white" Dutch people and English-speaking foreign residents; it is compatible with an officially desired goal of urban redevelopment and market-led gentrification. The other kind of globalization, marked by a concentration of low-status stores and cafés owned by transnational migrants and ethnic minorities, leads urban officials to fear "ghettoization." In its place, the urban planners bring redevelopment and social "diversity," as they say, in a process of state-led gentrification.

Utrechtsestraat, an upscale shopping street in the Canal Belt, and Javastraat, a downscale shopping street east of the city center, offer dramatic images of these two paths (see Figure 1).

Two Shopping Streets: Utrechtsestraat and Javastraat

At first glance, both Utrechtsestraat and Javastraat look like typical local shopping streets in Amsterdam. Each is fairly short and narrow, bordered on either end by a small square with green space. Both streets are filled with four- and five-story townhouses, most with red brick façades and white wooden trim, and each house has a storefront on the ground floor, adding up to about 130 shops on each street. Unusual even in Amsterdam, both streets host only a few chain stores.

But there the resemblance ends, for Javastraat is a low-price, "bread and butter" street in a working class neighborhood where immigrants, elderly Dutch, but increasingly also students and young professionals live, while Utrechtsestraat is a fairly expensive street of restaurants, cafés and bars, and distinctive clothing,

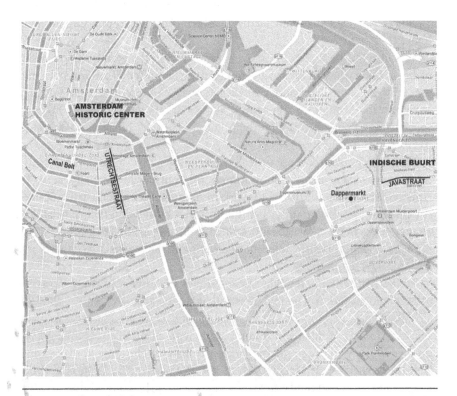

Figure 1 Map of Amsterdam, Showing Utrechtsestraat and Javastraat

Source: Google Maps, adapted by Sebastian Villamizar-Santamaria.

design, and food shops, surrounded by elegantly restored old canal houses and offices for financial institutions and new media firms (see Figures 2 and 3). Underlining these differences, commercial rents on Utrechtsestraat are more than twice as high as on Javastraat (DTZ Zadelhoff 2013).

Figure 2   Streetscape, Utrechtsestraat
Source: Photo by Richard Rosen.

Figure 3   Streetscape, Javastraat
Source: Photo by Sharon Zukin.

Javastraat is home to a concentration of ethnic shops and immigrant shop-keepers, mostly from the Global South. Utrechtsestraat, by contrast, has very few immigrant-owned stores and no businesses that are ethnically marked except for restaurants that serve foreign cuisines. These restaurants offer a choice of Italian, Indonesian, Indian, Thai, Mexican, and Japanese cuisines, but, significantly, omit Turkish, Moroccan, and Surinamese food, representing the city's three largest ethnic minorities. As this quick sketch suggests, both Javastraat and Utrechtsestraat are in their own way "cosmopolitan," but they represent different kinds of social and cultural capital.

For this reason, the two shopping streets generate different responses by the city government: benign neglect, in the case of Utrechtsestraat, and active displacement, for Javastraat. During the past few years, Utrechtsestraat has seen the opening of more high-priced, cutting-edge clothing boutiques and shoe stores. Meanwhile, the city government has rebuilt the streetscape of Javastraat, mandated the renovation of storefronts, and actively recruited Dutch chains and new retail entrepreneurs.

The paths of these two streets meet in a trendy style of décor and display that treats migrants and former colonies as an aesthetic inspiration. On Javastraat, a new bar is named for, and decorated in the spirit of, Walter Woodbury, a nineteenth-century British photographer who documented the Dutch East Indies, a region from which some of the street's shopkeepers come (see http://walterwoodburybar.nl). On Utrechtsestraat, the website of a new "concept store" selling "lifestyle" clothing and products geared to young men shows well-dressed models named "Omar" and "Pedro," a reference to the very migrants whom the street does not usually serve (see http://didato.nl).

How the paths of these two streets came to this curious point of convergence shows the power of the market on Utrechtsestraat and the power of the state on Javastraat.

Close Up: Utrechtsestraat

Utrechtsestraat has been an important radial street, or "street of access" to the city center, since the eighteenth century, as well as a major location for retail shops (Lesger 2007: 46). At one time a gate to the city stood at its southern end, and local legend says that wagon drivers on their way from Amsterdam to the city of Utrecht always stopped for a drink in the café which has been doing business on that street corner, under the ownership of the same family, since 1879. Today, a tram runs on a single track in the middle of the narrow street, linking the city's historic core to the southern business district and convention center.

Cultural Heritage

A strong sense of history is embodied in this street, from the architecture of the buildings and design of the shops, to the shape and scale of the surrounding streets. Located at the southeastern end of the Canal Belt, which was declared a UNESCO World Heritage Site in 2010, Utrechtsestraat is bisected by three of Amsterdam's best known canals. Many of the canal houses and warehouses in the neighborhood are designated historic monuments, as are the interiors of two shops on the street.

Though the oldest houses on Utrechtsestraat date back to the 1680s, a large number have been renovated over the years, and many show the aesthetic influence of the Art Nouveau period of the early 1900s. The overall look of the street is distinctive, but it also feels *gezellig* ("cozy"), a word that local residents use to describe it.

As this kind of embodied cultural capital might suggest, household incomes in the southern Canal Belt (Grachtengordel-Zuid) and the adjacent Weteringschans district are among the highest in the city (Bureau Onderzoek en Statistiek 2013: 101–3).[2] Store owners whom we interviewed describe their local customers as "captains of industry," "entrepreneurs who have big family businesses for several generations," and an "old bourgeoisie," meaning families who have owned canal houses for several generations. Business owners also refer to local residents as "intellectuals," by which they mean well-known media personalities, lawyers, writers, and professors.

According to store owners, all of their local customers share specific characteristics as consumers. They "want good quality but they don't want to spend money." They avoid conspicuous brand-name logos, and prefer the inconspicuous consumption of high-quality goods. As both a shoe store owner and the manager of a high-end audio store told us, their customers prefer the quality and cachet of European products to those made in Asia.

"Here," the president of the street association says, "you have a lot of high-quality entrepreneurs. We have the best butcher in Amsterdam, the best patisserie. We have Loekie's [delicatessen], which is well known for generations." He looks out the café window to the cookware shop across the street and nods in its direction. "Studio Bazar, everyone goes there no matter what it costs." He points down the block to Concerto: "One of the best record shops in the world."

Though store owners on Utrechtsestraat say that the majority of their customers are locals, expats and foreigners make up an important segment of their clientele. Most are North Americans, Britons, French, or Germans, who often work in transnational corporate offices in the city, as well as visitors who are staying at nearby hotels or who pass by on the tram on their way to and from the convention center. For this reason, most shopkeepers and employees on Utrechtsestraat speak English very well.

Just as there is little social class diversity, so there is little ethnic diversity among local residents. "This is a white neighborhood," says a real estate agent who has lived both on and near Utrechtsestraat since the 1960s. She has known only one family from Surinam and one from Morocco to live in the area during all these years. In fact, 63 percent of local residents are native-born (and presumably "white") Dutch. Among the others, 26 percent were born in Europe, North America, or Japan, and only 10 percent in "non-Western" countries (Bureau Onderzoek en Statistiek 2013: 25). Unlike in many other neighborhoods, there is practically no subsidized social housing.

From Prewar Prosperity to Postwar Decline

Utrechtsestraat seems to have always been both a local shopping street and a "destination" street of specialty stores (van Duren 1995). In the early 1900s, according to interviews with business owners who know the street's history, there were around fifteen shoe stores and almost a dozen butcher shops, as well as a well-known candy store, a piano and organ store, a crafts shop, an outfitter of hunting trips, and a shop that sold every kind of knife. By the 1950s, specialized stores for affluent residents outnumbered "bread and butter" shops for everyday needs.

Concerto, one of the first record stores in Amsterdam, opened on Utrechtsestraat in the early 1950s and soon developed a citywide reputation for its wide-ranging selection. Another specialized business, a kosher delicatessen, was opened in 1945 by a Dutch Jewish survivor of Nazi concentration camps who returned to Amsterdam after World War II. He received special permission from the city government to open the deli on Sundays; it drew shoppers to Utrechtsestraat to buy milk or fresh food on a day when nearly every other store in the city was required by law to close.

Other food shops were relegated to a side street, while "dirty" services like plumbers, bicycle mechanics, and garages clustered on another (unpublished historical data from Aart van Duren, 2010). Then as now, locations of specific types of shops were determined by the city government down to the level of which kind of business is permitted to operate in any given storefront. This is a narrower form of zoning than in other cities around the world.

Yet Utrechtsestraat declined during the 1960s, when many families left old housing in the city center for new homes in suburbs and smaller satellite cities. Like suburbanization in other regions of the world, newly developed residential areas attracted supermarkets and other large-scale retail entrepreneurs, and many families stopped shopping for their everyday needs in the center.

Shoppers were also anxious about the spread of illegal street prostitution and drug dealing from the red light district to the Canal Belt in the 1960s and 1970s. According to a store owner who lived above his family's shop at that time, at first the prostitutes began their workday after the stores closed at 6 p.m.

But gradually the street workers were younger and worked longer hours, the atmosphere grew more dangerous, and fewer shoppers wanted to come there.

Some businesses followed their customers. The oldest remaining shoe store, dating from the 1880s, opened a branch in the new southern district, in the city's first indoor shopping center. The kosher delicatessen also moved south. By the 1980s, the owner of the computer store claims, Utrechtsestraat looked like "second-hand city."

Despite the overall perception of decline, one area of growth was *horeca* businesses—*ho*tels, *re*staurants, and *ca*fés—for Utrechtsestraat begins at Rembrandtplein, a longtime center of popular theaters and movie houses, and is also near the Carré Theater. During the 1960s, when both tourism and Dutch disposable income began to grow, the number of restaurants and cafés rapidly rose, climbing from about 12 percent of all the storefronts on the street at the end of the 1960s to 30 percent in 1980 (see Figure 4).

Commercial and Residential Gentrification

Utrechtsestraat soon developed a reputation for trendy restaurants and design shops, which we now recognize as signs of imminent gentrification. Moreover, when Fred de Leeuw, a butcher who had sold cheap cuts of meat at wholesale prices to restaurants, decided to transform his business into an upscale shop and sell imported French truffles on the side, the media began to call the street

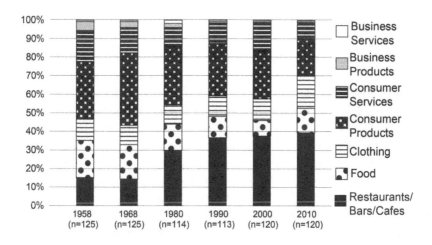

Figure 4 Types of Businesses on Utrechtsestraat, 1958–2010

Source: Research and Statistics Department, Municipality of Amsterdam; Aart van Duren; walking census. Graph constructed by Anke Hendriks and Sebastian Villamizar-Santamaria.

a foodie's paradise (van Duren 1995: 169). Yet at the same time, the number of vacant buildings rose, and Utrechtsestraat was termed a "graveyard" for retail businesses (van Duren 1995: 170; Arnoldussen 1996).

During the 1980s, local residents formed neighborhood watch committees to make note of license plate numbers on the cars of prostitutes' customers and report them to the police. At one point, they even tore paving stones out of the streets so "johns" could not drive through. This had an effect on the police, who by 1986 forced prostitutes and drug dealers to leave the area (Arnoldussen 2011).

Reducing crime made the area more attractive to middle class men and women, especially families with young children, who wanted to live in the heart of the city. They benefited from relatively low housing prices and zoning changes that favored residential conversion of office space, which encouraged them to buy old houses and warehouses that were no longer attractive as business locations. Home buyers also benefited from tax credits for restoring historic houses. These residential gentrifiers formed a growing market for new restaurants and upscale shops.

Though the variety of stores on Utrechtsestraat remained more or less the same, the number of restaurants and cafés continued to rise. By 1990, they occupied 40 percent of all the storefronts. The quality of goods and services also improved. For example, a pastry shop that had been doing business on the street since the 1970s changed its name from the quite normal Banketbakkerij ("cake bakery") to the more exclusive-sounding Patisserie Kuyt, and won an award in 2006 as the best patisserie in Amsterdam. By 2010, two housewares shops catered to foodies, along with the expensive butcher, a "normal" butcher, two cheese stores, a fresh fish shop, and a fruit and vegetable store that local residents called "the jeweler of greengrocers," for both the quality and prices of the products.

Although Utrechtsestraat maintains a small assortment of local shops for daily needs, like a dry cleaner, hardware store, and shop selling personal care products, it increasingly features "destination" boutiques for clothing, shoes, and jewelry. Concerto occupies five storefronts and offers a large, specialized selection of music, books, and high-end audio equipment, but it has lost many customers to online music downloads. To maintain its appeal, Concerto now sponsors live performances on Sundays and operates a café inside the store.

The changing array in businesses in the street reflects the upward trend in property values in the neighborhood, the continued high income level of local residents, and the strong presence of financial and new media firms. Yet because all the shops are small, the interiors encourage conversations between shopkeepers and customers.

The long counter in Loekie's delicatessen is a center of sociability during the day, especially at lunchtime, when the shop does a good business in takeout sandwiches, each prepared individually as customers order them. The cafés and small corner bars are also social nodes, especially on Saturday afternoons before shops close and again at nighttime. In fact, one-third of all the street corners on Utrechtsestraat are occupied by bars, which is an unusually high concentration for any city. Locals describe each bar as catering to a specific clientele, from yuppies to intellectuals and even cross-dressers, and all of this specialization confirms the street as a nighttime destination.

Specialization is accentuated by the visible absence of chain stores. Less than 10 percent of the storefronts on Utrechtsestraat are occupied by chains, and most of these look more like individually owned shops. In 2008, however, the son of a family that had owned a modern furniture store on the street for years closed the shop and rented the double storefront to the U.S.-based clothing chain American Apparel. Two years later, a branch of Marqt, a regional organic foods supermarket chain, opened at the northern end of the street, and, despite complaints from some local residents, a branch of Albert Heijn, the Dutch supermarket chain, opened around the corner at the southern end.

Today, Utrechtsestraat remains cozy, convenient, and inconspicuously luxurious: an "urban village" for the cosmopolitan middle class.

An Urban Village for the Cosmopolitan Middle Class

For locals, the street's most important feature continues to be its sociability. From the manager of the cheese store to the owner of the newest café, everyone describes it as an "urban village." Except in bars, which tend to have absentee owners, most shopkeepers work in their shop every day. So they know their customers and know each other, too. "It's like a little village in the big city," the new café owner says. "Everyone knows each other. When we opened, the other owners brought us flowers and cakes, and wished us good luck."

Business owners buy supplies from each other if they can. But hardly any of them live above the shop or even in the neighborhood. Most do not own their building. But even if they did, because many buildings lack a separate entrance to the upper floors, it is illegal to use them for residence. The street's sociability, then, depends on daytime activities, when store owners and employees are highly visible through the plate glass window, usually working behind the counter, and on the nighttime attraction of the bars.

The presence of most business owners and the small size of stores create an intimacy between shopkeepers and customers. This is especially true in shops where owners and employees chat with customers across traditional counters. Food stores play a key role in constructing the urban village effect, not only because of their counter and small size, but also because the same owners and employees serve the same customers day after day and year after year. Both intimacy and continuity get special reinforcement in the cheese shop, delicatessen, and greengrocer, where the owners come from families who have worked in the same business for two and three generations (see "Shopkeepers' Stories" section on Loekie's Delicatessen, page 112 below).

Indeed, the ecosystem on Utrechtsestraat has a remarkable stability. A bar near the old gate to the city, a shoe store, and a bakery all date back to the 1880s. More important, an extraordinary one-half of all businesses on the street have been there at least 20 years, one-third have been there more than 30 years, and one-quarter have been on Utrechtsestraat since at least the 1970s.

Nearly all the store owners rent their space and complain that rents are rising. Nevertheless, longevity tells us that business is good and rents are bearable. Commercial leases are written for a longer term than currently in cities like New York—for five years, in Amsterdam—and include an automatic five-year renewal with rent rises limited to the inflation rate.

If we look for continuity in the same *types* of stores, the ecosystem's stability is even more striking. At least three-fourths of the storefronts have been occupied by the *same* business or same *type* of business since at least the 1990s. More than half the storefronts have been occupied by the same type of business since at least the 1980s, and more than a third since at least the 1970s. As the manager of Concerto's audio shop says, "When you come on your bike and enter the street, you feel it is *your* street." (See Figure 5.)

The longevity of the stores complements the historic architecture of the street, with both emphasizing a sense of cultural continuity. As the owner of one of the "coffee shops" says, people like Utrechtsestraat "because it reminds them of back in the old days. Amsterdam has changed completely, but this street still has the former atmosphere ... For years and years the area looks the same." The shopping street, in short, is a constant reminder of the city's cultural heritage (Zukin 2012).

But the sense of belonging in Utrechtsestraat refers not only to cultural heritage, the "urban village" effect, and the long history of many businesses.

Longevity of Stores in 2010: Same Name OR Same Type

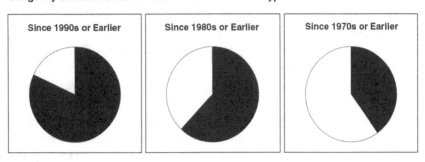

Figure 5 Stores on Utrechtsestraat Tend to Have Long Lives

Source: Research and Statistics Department, Municipality of Amsterdam; Aart van Duren. Graph constructed by Anke Hendriks.

The ecosystem of shoppers and shops is founded on, and reinforces, an absence of social and ethnic diversity. Only a handful of business owners come from ethnic minorities, and some of them run high-status shops that are indistinguishable from all the others (see "Shopkeepers' Stories" section on Kenza Hair Salon, page 115 below). Although this homogeneity reflects the demographics of the surrounding neighborhood, it does not represent the city as a whole.

Nevertheless, Utrechtsestraat impresses both neighbors and government officials as a street that is doing well. For this reason, the street repairs that began in 2009 caused a real crisis in relations with the local state.

Street Repairs Mobilize Business Owners

Shopkeepers' satisfaction with life on Utrechtsestraat was severely strained when the city government began major street and bridge repairs in 2009. Not only was the tram line moved for an indefinite period of time, but the construction of barriers and signs made it appear as though the entire street were closed, which convinced customers to stay away. This situation, coinciding with the financial crisis that began in 2008, caused a drastic loss of business. Store owners bitterly complained, but felt that no one at City Hall was listening.

To deal with the street repairs, a few business owners revived the street association, a traditional shopkeepers' organization which existed in the past but had become inactive. One of them suggested that the association hire Nel de Jager, an activist street manager who had brought new life to Haarlemmerstraat, a local shopping street at the northwestern end of the Canal Belt, where during the 1970s derelict buildings, slated for demolition, had housed drug dealers and squatters. Initially working on her own, de Jager recruited artisans and small retail entrepreneurs who wanted to open specialty shops. She persuaded building owners to rent them space at reduced rents, and gradually, Haarlemmerstraat developed into a lively and desirable location. When the merchants formed a street association, they hired de Jager as the manager. Half the salary for this position is paid by the street association, and half by the city government.

On Utrechtsestraat, there was no need to recruit new businesses. But when de Jager signed on to work as the street manager, store owners gained an experienced advocate to argue their case to the city government.[3]

For the next two years, the executive committee of the street association devoted all of its meetings to discussing the street repairs and the impact of the tram's absence on business. Working through the street manager, they were able to get the city government to remove the barriers. Then they demanded to know the precise timetable of repairs, and pressed the city's transportation office to outline alternative strategies that would both reduce interruptions to auto traffic and bring back the tram.

At the same time, de Jager performed the same role she had developed on

Haarlemmerstraat. She acted as a go-between, looking for suitable businesses to rent vacant storefronts and advising their owners on how to adapt to local conditions and apply for government subsidies for restoring historic buildings. She also organized street festivals, sometimes in cooperation with outside cultural organizations.

Very important for the image of Utrechtsestraat, de Jager proposed promoting the street as a *klimaatstraat* ("climate street"), and arranged meetings with representatives of local utility companies to reduce store owners' bills for electricity and trash removal and install more cost-effective street lights. For these efforts Utrechtsestraat won an award in 2011 as best *klimaatstraat* from *Time Out Amsterdam* magazine.

Most important, the street manager's continued advocacy caused the city government to change the schedule for street and bridge repairs, and the tram returned ahead of plan, in 2012. Meanwhile, de Jager suggested to the executive committee of the street association that they take advantage of a Dutch pilot program allowing for the formation of business improvement districts (BIZ). In the fall of 2011, a majority of business owners on Utrechtsestraat voted to do so.

The main benefit was financial. Though most store owners seemed satisfied with the street association's efforts, not everyone had been paying the charges needed to support the annual two months of holiday lights and special events. In a BIZ, as in business improvement districts elsewhere, dues must be paid in the form of a mandatory tax by every business using property on the street. Unlike in New York, building owners neither belong to the BIZ nor pay a tax to support it.

Changing the street association into a BIZ created a sustainable financial base for all "activities that contribute to the livability, safety, regional planning and other public interests in the public space of the BIZ zone" (Staatsblad van het Koninkrijk der Nederlanden 2009). Unlike in the U.S., a BIZ is not permitted to hire its own street cleaners or private security guards. But the mandatory tax would pay for holiday lights, the street's website, and special promotions. To celebrate the tram's return, the BIZ held a festival with children's games and stands operated by local shops.

After the crisis of street repairs was resolved, business owners turned their attention to two different concerns they were powerless to combat: rising rents and a disproportionate number of restaurants, cafés, and clothing stores. On the one hand, new building owners, including a small number of corporate investors, showed more interest in raising rents than in keeping old tenants, and this attitude was bound to lead to more frozen yogurt shops—the latest trend, restaurants, and expensive boutiques. On the other hand, trendy stores found rents on Utrechtsestraat to be lower than in one or two other popular areas, a cost advantage that would attract even more of them.

The opening of a Mark by Mark Jacobs clothing boutique reflected both the street's commercial success and the risks of what Nel de Jager called "too much, too high." There were too many stores with high prices that most Amsterdammers could not afford, and she feared that the lack of social diversity would not sustain the street's appeal.

Close Up: Javastraat

Like Utrechtsestraat, Javastraat also developed an "urban village" feel, but it differed greatly in the social status, ethnic identity, and cultural capital of its stakeholders.

A Neighborhood in Transition

Javastraat has been the main shopping street in the Indische Buurt (the "Indies Neighborhood," named for the former Dutch East Indian colonies) since it was built in the early 1900s. The neighborhood's development reflected the opening of new harbors and canals at that time, and the accompanying dramatic expansion of manufacturing and transportation industries. As on the Lower East Side of Manhattan several decades earlier, thousands of quickly constructed tenements in the Indische Buurt offered affordable housing to Amsterdam's rapidly growing working class. But conditions changed in the 1960s and 1970s, again as in New York, when the region's manufacturing industries began to decline, and the city's older districts, with aging and dilapidated housing, lost their appeal.

Many residents of the Indische Buurt, including native-born Dutch skilled workers, small business owners, office clerks, and government employees, moved to the suburbs. Large numbers of unskilled immigrants from developing countries—Turkey and Morocco in particular—who had come to the Netherlands for factory work, moved into the apartments they left behind. However, houses in the Indische Buurt had never been well built. By the 1970s, many were in a severe state of disrepair (Heijdra 2000). And all residents who worked in manufacturing faced an imminent loss of jobs.

Following the orientation of urban policy in the Netherlands during those years—"building for the neighborhood"— urban renewal programs aimed to renovate or replace run-down tenements on behalf of the existing, working class population (Van der Pennen and Wuertz 1985; Smit 1991). In line with the dominant political ideology, any proposal to cater to the middle classes met with fierce opposition from local residents and their organizations (Heijdra 2000).

Yet preserving the working-class character of the neighborhood, once the dream of community activists and local politicians, is considered to be highly problematic today.

More than thirty years after the first round of renovations, the concentration of low-cost housing in the Indische Buurt became the target of new policies for change. This time, the city government aimed to replace working-class residents, who by now were mainly immigrants, with a middle-class population. Strategies were created to raise rents to market levels and convert rental housing to owner-occupancy.[4]

Privatization policies on the national level laid the groundwork for these goals. During the 1990s, the housing associations that manage Amsterdam's "social" housing stock—and therefore own most real estate in the area around Javastraat, but practically none near Utrechtsestraat—were spun off by the government and privatized. Since then, subsidized housing has been provided and maintained by a small number of nonprofit organizations, which must pay for their operations without government subsidies. They generate income by selling or renting out part of their assets, the traditional social housing stock, at market prices.

Typically, this works best for them in centrally located, architecturally attractive buildings which can be gentrified. But in light of high housing prices in the center of the city, and the recent transformation of abandoned docklands immediately to the north into living lofts, artists' studios, and cultural facilities, the area around Javastraat began to look ripe for gentrification by the "creative class."

Before the 1990s, the strong presence of housing associations would have prevented or at least slowed the gentrification process. But today, a consortium of housing associations and the local government is the main driver promoting the transition of the Indische Buurt into a mixed-income neighborhood. They follow the "mixing" policies of both the national and local governments (Stadsdeel Zeeburg 2007), which target concentrations of poverty and disadvantage for an increase in social "diversity," and the "creative city" policies of the city of Amsterdam (Peck 2012).

Unlike efforts to increase diversity in the U.S., "mixing" policies bring an influx of middle-class, "white" Dutch residents and "creative" businesses into areas with strong concentrations of ethnic minorities and immigrant-owned shops. These developments have had profound implications for the multicultural urban village that Javastraat had become.

A Multicultural Urban Village

Entered from the west, the side nearest the city center, Javastraat stands out for its multi-colored store awnings and abundant sidewalk displays of fruit and vegetables. In many shop windows, prominent signs listing low prices suggest that undercutting the distant supermarket and, more important, nearby competitors is a common business practice.

At first glance, the street almost seems to consist entirely of food stores and deli shops. But in fact, the greengrocers and small delis are joined by shops selling everyday products like housewares, cosmetics, clothing, accessories, books, films, and music. Prices are low in most of the stores, and some of them, with signs for halal meat or Bollywood films, clearly aim to attract an ethnic or immigrant clientele. Conforming to Javastraat's reputation for bargain shopping, many food stores sell products in bulk. Compared to the store displays on Utrechtsestraat, their stock looks messy and disordered. On busy days, the cluttered chaos gives the street a bustling atmosphere like that of a street market. But when there are fewer people on the street, the clutter gives the impression of a shopping street in decline.

In the 1960s, Javastraat had shops selling more expensive clothing, shoes, and jewelry. Some local residents and shopkeepers remember this as a time when Javastraat was an upscale destination. By the 1980s, however, the street was mostly dedicated to products for everyday needs, and it has kept this functional character.

As on Utrechtsestraat, the share of different retail sectors on Javastraat has remained relatively constant, but in contrast to the upscale street, shops have often changed hands (see Figure 6). In fact, every ten years since 1980, a majority of storefronts have changed both ownership and type of business. Only three stores on Javastraat today already existed under the same name and ownership in 1980: two opticians and a pet shop.

Immigrants gradually became more established in Javastraat as both shop owners and customers. In 1980, as far as we can see from owners' names, a "non-Dutch" ethnic presence was limited to two Chinese restaurants, a

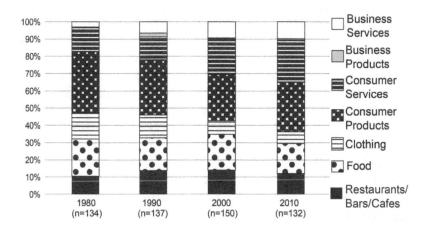

Figure 6   Types of Businesses on Javastraat, 1980–2010

Source: Research and Statistics Department, Municipality of Amsterdam; walking census. Graph constructed by Anke Hendriks and Sebastian Villamizar-Santamaria.

grillroom, kebab shop, Israeli snack bar, and two halal butchers, out of 199 shops operating at that time.

Since then, the majority of shops on Javastraat have operated under non-Dutch (and non-European) names. Consistent with the residential demographics of the Indische Buurt, Moroccan and Turkish shops are prominent. Some of these target a non-Dutch ethnic clientele, such as a Moroccan coffee house, an Islamic book and clothing store, and a shop with wigs, braids, and products for black hair (historically, native-born Dutch are predominantly blond). Others target a mixed clientele.

The transition towards ethnic shops unfolded gradually, but became especially noticeable for longtime residents with the disappearance of the last native-born Dutch butcher between 2000 and 2010. This left Turkish and Moroccan, both Muslim, butchers selling halal meat as the only option. Furthermore, with the closure of the last remaining cheese shop and liquor store after 2010, two types of food stores that are unlikely to be opened by ethnic, or at least Muslim, shopkeepers disappeared from the street entirely. Though ethnic diversity in the shops is increasingly felt by longtime residents, "Dutch" character is not, and gentrification is becoming more tangible, too.

In 2010, a small shop in Javastraat with a tiny supply of old vinyl records and a few items of vintage clothing was a rare sign of hipster gentrification. By 2013, however, new bars and cafés were drawing customers (Ernst and Doucet 2014), and an ambitious restaurant with a trendy industrial aesthetic and a name bound to repel Muslims—De Wilde Zwijnen (The Wild Boar)—was getting good write-ups in local and global media, including the *New York Times* and TripAdvisor.com.

At the western end of the street, in a store called DIV (for "diversity," the white Dutch owner told us), men's clothes hang from scaffolding tubes and rough black elastic wires that trace a zigzag path through the store. Prices for jeans are high, including some made of Japanese selvedge denim. A few doors away, a large new bar, the Walter Woodbury (mentioned earlier), has drawn complaints from residents because of noise made by its patrons late at night (Parool 2014). Across the street, customers sip cappuccinos at tables placed on the sidewalk in front of the Bedford–Stuyvesant Café, named—ironically to us researchers—for the predominantly black Brooklyn neighborhood we studied in New York.

Around the corner, on Javaplein, a new designer furniture outlet and fitness club rent space in an award-winning new building. This square has been completely renovated into a playground and green space with fountains, driving away the immigrant teenagers who used to gather there. As marked by a plaque in the new red-brick paving, the same used in the historic city center, the renovation was funded by a grant from the European Union.

On sunny days, the square is taken over by sidewalk cafés where shoppers and local residents have lunch or a snack, very much like on Utrechtsestraat and other shopping streets in the city center. A point of attraction is the red neon logo of the Coffee Company, a Dutch chain comparable to Starbucks that is increasingly regarded by Amsterdammers as the ultimate sign of gentrification. Nearby, tech startups have been recruited as commercial tenants of renovated buildings.

The public at the sidewalk cafés is mixed, but differs from the rest of the street. Though Javastraat still has many female shoppers wearing the long dresses and head coverings typical of observant Muslim women, and shoppers who are visibly members of ethnic minorities, the new restaurants and cafés on Javaplein are mostly frequented by native-born, white Dutch. This division is even more marked in the evenings, when a more affluent public from all over the city is drawn to the restaurants that have opened in the past few years.

From this point of view, Javastraat is beginning to look like Utrechtsestraat. But its commercial gentrification has taken place about twenty years after Utrechtsestraat's, and in a different economic and public policy context.

State-Led Commercial Gentrification

Although residential gentrification has swept through Amsterdam as it has done in other cities of the world, the city government and housing associations that used to offer subsidized apartments decided to make it happen faster in the Indische Buurt by changing the retail landscape. They reasoned that highly educated professionals would more likely buy apartments there if they could be surrounded by trendy restaurants and cafés.

As the main shopping street in this neighborhood, Javastraat became a major target of their intervention, beginning in 2007. Many of the prewar brick townhouses now sport freshly sandblasted façades with neatly painted, white or off-white wooden trim, while signs announce, "*Te Koop. Luxe appartementen*" ("For Sale. Luxury apartments"). The city planning department, without asking store owners, designed a new streetscape, reducing parking space for cars but increasing it for bicycles, tearing out trees, and installing new planters. Like the upper floors, storefronts have been renovated and freshly painted, and many windows show "For Rent" signs, often specifying the space is only for "*horeca*" (restaurants and cafés).

Housing associations selling property online emphasize the "new" Javastraat's shops and cafés in order to attract residents who are quite different from ethnic shoppers:

> The Indische Buurt in the East of Amsterdam has been rapidly increasing in popularity over the last couple of years. This up-and-coming neighborhood ... has been referred to as an "East Village" by lifestyle magazines as well as by locals. The apartments are located on Javastraat with all its shops and dining venues. You're right near Studio K [a cinema,

music club, and café], Café het Badhuis [a restaurant in a former public bathhouse], restaurant De Wilde Zwijnen and the Coffee Company (Funda undated).

As the street manager, Marcia van de Hart, says, "New residents are important in this process [of changing the neighborhood]. They demand [commercial] diversity. If we would not make it easy for them, we would lose them to other shopping areas. In that case, the residents that are able to make a difference in the neighborhood go elsewhere."

The street manager makes it clear that "diversity" should not be understood in the sense of the multicultural urban village that Javastraat had become. Although "diversity" in New York usually refers to the cultural value added to a shopping street by ethnic restaurants and immigrant-owned stores, the same term in Amsterdam signifies the cultural value added by the ABC's of gentrification: art galleries, boutiques, and cafés.

Unlike the former street manager (and also the current BIZ director) on Utrechtsestraat, the street manager on Javastraat works for the city government. Rather than advocating for the merchants, she coordinates actions taken by the government and housing associations and communicates them down to the shopkeepers. She is specifically charged with using public policies to promote Javastraat's upgrading—which, in contrast to the administrative decisions that control it, she describes as "an organic process." She is also the eyes and ears of the city government, responsible for passing information about the street upward to the urban planners.

The local government and housing associations have three main ways to promote change on the street: by direct regulation, indirect regulation, and property ownership. On Javastraat, commercial property is nearly all owned by private landlords. This leaves direct regulation through zoning laws, and indirect regulation through special, targeted rules, as the local government's tools for redevelopment.

Zoning plans in Amsterdam determine which kinds of businesses can legally be established in any street. Although the goal is to limit competition, planners manipulated the zoning rules on Javastraat to increase the number of restaurants and cafés that they believed would appeal to "creative" and middle-class residents. They also shifted *horeca* permits between lots to enable cafés and restaurants to cluster together in a critical mass. As both street managers told us, urban planners pay much closer attention to the kinds of stores operating on Javastraat than they do on Utrechtsestraat.

One reason for the closer attention to store types on Javastraat is to prevent criminal activities ranging from money laundering to illegal drug and weapon sales, which motivated a wave of shop closures by the city government in the late 2000s. During that crackdown, a launderette, a video store and a coffee house were closed.

Since then, the Javastraat street manager has cooperated closely with the police, who still monitor several stores in the street and predict another wave of shop closures. Nevertheless, there is little street crime, perhaps because those who control the other criminal trades don't want to call attention to their illegal businesses.

Call centers, where customers make low-cost, long-distance phone calls and use the Internet, have drawn particular scrutiny throughout Amsterdam, but especially in the Indische Buurt, where they mostly cater to very poor, first-generation immigrants. Rightly or not, they are associated with informal economic practices and criminal activities. On Javastraat the few call centers that remain are tolerated as long as they last, while zoning plans prevent any new ones from opening.

But because zoning plans are reviewed only once every ten years in a time-consuming, public process, government officials try to produce rapid change in the retail landscape by indirect regulation. On Javastraat, they targeted the aesthetic look of the space.

In the early 2000s, the city government demanded that landlords on Javastraat renovate the structure, façade, and general appearance of all the shops. They offered subsidies for making renovations, as long as the shopkeeper and/or building owner also paid a share. Some new rules required building owners and shopkeepers to use only paint colors that are compatible with official plans—mainly, the white and off-white, black, and dark green that are used in the historic city center.

The city government's main strategy, however, is to go out and recruit "good quality" businesses which will attract the desired "diversity" of residents. This may take some time. In that case, the street manager tries to persuade building owners to accept the lack of rent "while they look for a sound entrepreneur" who fits the "long-term vision" for the street.

But the delay may take too long for some landlords. So the city government asked the housing associations to buy up some of the commercial property on Javastraat, renovate the storefronts, and rent them out for reasonable (i.e. subsidized) prices to "approved" shopkeepers. The exact criteria for which businesses are "approved," and which are not, are largely taken for granted and rarely commented on. Yet observations of new businesses on the street, and interviews with new business owners, suggest that ethnicity and cultural capital play a huge role.

Aesthetic, Not Ethnic Diversity

The street manager expresses the official view: the ethnic grocery stores are financially marginal, they all look alike, and there are too many of them. She complains that, although some ethnic grocers took advantage of subsidies to upgrade their stores, they used this money to expand their selling area without making it look like the kind of specialty shop that would attract middle-class shoppers.

According to the city government's vision, the ethnic food stores should display their "exotic" wares, such as dried dates, in the front of the store and in sidewalk displays. This would make Javastraat look like a "Mediterranean shopping street." Instead, the food store owners display everyday products on the street. "They often direct [these displays] to their own target group, the ethnic residents," the street manager complains. "They do not appeal to the rest of the neighborhood … They always display their shiny cucumbers and tomatoes in front of their stores. It always seems as if all they have is cucumbers and tomatoes."

Redevelopment of Javastraat is not aimed at eliminating an ethnic presence, but at helping ethnic entrepreneurs to "take their businesses to the next level," apparently, by catering to the tastes and exotic fantasies of a native-Dutch, middle-class public. This suggests a tension between the dream of a "Mediterranean" market on Javastraat and the success of Dappermarkt, a nearby street market selling inexpensive ethnic products, which in 2007 was judged both the best market in the Netherlands and one of the top ten shopping streets in the world by National Geographic Traveler (Amsterdam.info undated).

The policy aim on Javastraat is not to decrease the number of food stores. Rather, the ethnic grocers are expected to "professionalize" their shops so they will be more attractive to gentrifiers. Yet when an ethnic greengrocer adapts to new, native-Dutch residents, and offers products oriented to their tastes, the small number of new customers whom they gain may not make up for the loss of old customers (see "Shopkeepers' Stories" section on Tigris & Eufraat, page 113).

The bias that underlies Javastraat's redevelopment is revealed by two shopkeepers who had considerable trouble renting storefronts from a housing association. One is the owner of a vintage clothing store, who says it took her a long time to convince the housing association that she was going to open a "quality vintage" shop rather than a second-hand "junk store." The other is the Moroccan owner of an upscale café, who says that, despite making several phone calls, he could not get an appointment with anyone from the housing association to discuss renting a storefront on Javaplein. He was only able to make an appointment after he asked his wife, who is also Moroccan but speaks Dutch without an accent, to phone the housing association for him.

"Desirable" Shopkeepers

By contrast, business owners who told us they were invited to locate on Javastraat include the owners of trendy restaurants and cafés, and of the Coffee Company chain. Some of them lived in the neighborhood already, and wanted to open a business that would cater to people like themselves, new residents who work in the arts and "creative" occupations (see "Shopkeepers' Stories" section on Comfort Caffe, page 116 below). Many claim that, like new retail entrepreneurs in gentrifying neighborhoods in other cities (Zukin 2010), they started their business on Javastraat "to do something for the neighborhood."

Yet in stark contrast to the ethnic business owners, these "desirable" shopkeepers benefit from the official revitalization plan. Though the housing associations try to entice them by offering lower than usual rents, the required renovations have almost doubled the rents for most longtime shopkeepers. An extra advantage for the new shopkeepers is that they usually move into property that has been empty for some time. In this way, they avoid high takeover costs, especially for restaurants and cafés, where new owners normally pay the previous owner a fee for "goodwill."

As on Utrechtsestraat, and for the same reasons, the business owners on Javastraat recently voted to turn their street association into a BIZ. The chair of the committee to form the BIZ was an established shopkeeper who had also chaired the previous street association, but the owners of the Bedford–Stuyvesant café and of a fairly new ice cream shop were members of the board as well.

The new store owners have a vision for Javastraat that is closer to that of the city government than the longtime ethnic shopkeepers. The owner of the Coffee Company says, for example, that he would be interested in organizing with other shopkeepers in partnership with the local government to "invest some money to remove all these trees and put in some real trees." This is remarkable, considering that recent renovations of the streetscape—undertaken by the city government without consulting the ethnic shopkeepers—involved the removal of many large, old trees.

But what kind of commitment do the new retail entrepreneurs really have? Rather than showing patience with small trees that will grow into bigger ones over time, they would like to see them replaced by bigger trees from elsewhere. This type of confidence among new shopkeepers contrasts with the disbelief among longtime ethnic business owners that they can lobby the city government to defend their interests, and stay on Javastraat for a long time.

Global Model or Local Ecosystem?

Today, both Javastraat and Utrechtsestraat are heading toward a global model of the city in which revitalized local shopping streets play a highly visible role. But on Javastraat, this model disrupts the fragile ecosystem of the urban village that developed over time, based on trust between shopkeepers and customers from ethnic minorities and the sense of belonging that they share. The global model also disrupts the upscale ecosystem of Utrechtsestraat, where rising rents and continued increases in boutiques and retail chains leave no place for "bread and butter" stores that build a sense of community.

Most important, the contrast between state policies toward Javastraat and Utrechtsestraat shows how a city government can manipulate "market forces" to shape commercial as well as residential gentrification. In Javastraat, this has not won universal approval. A recent newspaper article says that the official ideology of social integration has backfired because it promotes gentrification as the only path to neighborhood improvement, even if it means that most current shopkeepers will gradually disappear. However, as Auke Kok observes:

> That Javastraat should become more "white" is, surprisingly, what many of the stakeholders agree on. Shopkeepers with highly divergent ethnic backgrounds who do not know much—and perhaps do not understand anything—about each other, share the same desire for quality. And "quality" seems almost synonymous with native Dutch (Kok 2014, emphasis added).

This brings us back to Utrechtsestraat, where "quality" shops, intentionally or not, reinforce, and are reinforced by, a lack of ethnic diversity. Both Utrechtsestraat and Javastraat are "cosmopolitan" urban villages. But if market forces on the one street and government policies on the other continue as they have done so far, they risk making "diversity" into an exotic attraction. On U'straat, shopkeepers' narratives show complementary strategies of tradition and innovation; on J'straat, adaptation and gentrification.

Shopkeepers' Stories

Loekie's Delicatessen, Utrechtsestraat: Tradition

Sjoerd, a tall, ruddy-faced man in his late forties, is a member of the family that has owned and operated Loekie's delicatessen on Utrechtsestraat since 1945. His family has been merchants for generations on both his mother's and father's sides. His grandfather, who migrated to Amsterdam from an eastern region of the Netherlands, had four brothers, and they all started food businesses in the city.

Sjoerd's father was born in 1908. He opened his first business in 1926, and bought this store, which had been a food shop owned by Jews deported during World War II, in 1945, after the war. "He *was* the business," Sjoerd exclaims. "When he came into the shop he knew exactly what was happening. He was a classical, old-fashioned business man; he wore a hat in the shop. He was Roman Catholic, he went to church, and he was warm-hearted."

By 1945, when Sjoerd's family opened the delicatessen on Utrechtsestraat, his father owned six stores. "In one shop he sold cheese, in others, other food products. You can still see the words for coffee roasters and peanut butter on the big building that he owned on Prins Hendrikkade, in an old port area of the city. There were about 40 people working for him in all the stores."

The food business continued to grow during the 1960s, but problems arose because of taxes, the pension system, and competition from supermarkets. When Sjoerd's father died in 1971, "because of stress," the family sold four stores to straighten out their finances. Gradually, they raised the quality of the products that they handled. "You sell less but the revenue is greater," Sjoerd says.

During the 1980s, Loekie's prospered by catering lunches at offices of the Netherlandse Bank and ABN Amro Bank, at the two ends of the street, and selling sandwiches to their employees. Each had about 1,000 men and women working there. "In those days I couldn't close the store at 1 or 2 p.m. [the way we do on Wednesdays now], we had so many customers for lunchtime, from 11 [a.m.] to 2 [p.m.]. Gradually those workers went away, and so did the big offices." In fact, Amro moved its headquarters to a new office park in the south of the city, and it took a while for gentrifiers to move in. "There was another high point from 1995 to 2000 when the yuppies came, and since 2000 the business has stabilized."

The clientele "now is a mix, banks and all different businesses. A lot of young people [in the neighborhood] are starting their own business and they want good sandwiches from us." Loekie's also sells to tourists and expats, and is written up in online shopping guides.

"We have 12 to 14 employees in all, including three brothers and a sister." But when we ask whether the family thinks about opening the shop on Sundays, Sjoerd laughs. "You can't work 24/7," he says. "People don't work two jobs. I don't want that for me or for those who work for me. My father was born in 1908 when workers pleaded for a two-week vacation. They finally got that, and that's why we don't open on Sundays or in the evening. In big companies they are only responsible to their stockholders."

Tigris & Eufraat, Javastraat: Adaptation

Tigris & Eufraat, named after the rivers that flow through Turkey, Syria, and Iraq, is a greengrocer's shop that is owned and run by Youssef and his father. The family has owned the store since 2003. Youssef was born in Iraq and moved to the Netherlands when he was 8 or 9 years old. His father had left Iraq as a refugee years before that, when Youssef was still very young. Even after moving to the Netherlands, Youssef felt as though he did not know his father at all, since he had been absent for most of his childhood.

Youssef's father bought the shop from the previous owner. This was a rather arbitrary move, since he did not have a history in trading or groceries at all. He used to be a welder when he lived in Iraq and worked at the flower market when he moved to Amsterdam. However, he wished to be a business owner and got a tip about the store coming up for sale:

> Someone visited my dad and said, like, there's this shop, a vegetable shop, it's an Iraqi vegetable shop, and since you're Iraqi, perhaps it would be interesting for you to start your own business ... and then he came here, he took the risk. Yeah, but he didn't really choose [the kind of business it was]—it could have been a restaurant as well, it could have been a bakery.

Youssef had not really been planning to join his father's business, but ended up there when he was about to drop out of high school because he was, he says, more interested in girls than in the books. After he failed exams to move up to the next level of schooling, his father really had it with him and more or less summoned Youssef to join his business. Two years ago, his father made Youssef the co-owner.

Tigris & Eufraat sells some typical Middle Eastern products, such as falafel and camel's milk, so it attracts many Central Asian customers. For this reason, the shop forms a critical mass with other ethnic stores on Javastraat, especially those owned by other Iraqis: "We've got an Iraqi restaurant down the street, we've got an Iraqi bakery down the street, and we have an Iraqi vegetable store. So whenever Iraqis visit the Javastraat, they visit the three of us."

But for other kinds of products, there is fierce competition with other green-grocers in the street. The others buy the same products from the wholesaler and try to sell them at lower prices. Nevertheless, Youssef claims to keep out of these competitions as much as possible:

> There used to be fistfights sometimes, but fortunately, we never had any arguments with our neighbors. But we had some neighbors, here and across the street and down the street—the Turks against the Turks. A Turkish shop and another Turkish shop, they started competing with each other, while my shop was still the most expensive one—they competed with each other and I still did better business than the both of them.

Tigris & Eufraat has recently been redecorated and expanded. The building underwent a thorough renovation, as a local government inspector had established that the foundation needed repairs. At the same time, the building next to the shop was vacant. The building owner did the necessary repairs, aided by a government subsidy. In return, the building owner charged rent for only one of the lots during the time of the renovation. This was a difficult time for Tigris & Eufraat, because they decided to stay in business during the renovations to maintain as many of their regular customers as possible. But they had to work around the renovations, doing without refrigeration, for example, for a long time.

Today, the business is recovering from the renovations, although in a harsh economic climate. Due to the financial crisis that began in 2008, Youssef claims, there are fewer customers in the street than there used to be. In his opinion, this is aggravated by the renovations of the streetscape, since there are fewer parking spaces for customers. Moreover, the building renovations brought stricter enforcement of the zoning rules. Tigris & Eufraat has received four fines because their sidewalk display extends beyond the permitted distance. Finally, the renovations of the apartments above the shops have brought new residents to the street. Although some of them shop at Tigris & Eufraat, they are generally single- or two-person households and therefore need fewer groceries than the families that used to live there.

However, Youssef has tried to adapt to the new residents by stocking new products. Some residents, for example, started to ask for parsnips. This is a northern-climate root vegetable that had fallen into disuse in the Netherlands, but became trendy after being promoted in a campaign for "forgotten vegetables":

> Dutch people are always asking for parsnips. So I said: "Alright, tomorrow I'll have parsnips." I went and got parsnips, so since that day I sell parsnips … Albert Heijn sells parsnips in bags, it's too much. At my place you can just take one, or as many as you like.

Kenza Hair Salon, Utrechtsestraat: Innovation

Kenza Hair is a small men's hair salon on Kerkstraat, just a few steps off Utrechtsestraat. Instead of the concert and theater posters that cover a wall near the entrance of many stores here, the salon displays autographed soccer balls and shirts, players' photos, and a few fashion and *voetbal* [soccer] magazines neatly arranged in a rack. The décor of the immaculate salon is black and white; the back wall is covered in turquoise tiles. In the mirror on a side wall, clients can watch the owner, Ricardo, create unique styles for a client sitting in one of the two barber's chairs. Light hip-hop and pop music play on the sound system. On Saturday afternoons the place is packed with clients.

Ricardo is in his thirties. He is of medium height, very trim, and wears a polo shirt, athletic shoes, and loose but not baggy jeans. He smiles readily and speaks English very well, which he says he learned from watching movies. Ricardo's skin is golden tan, suggesting both his mother's white Dutch origins and his father's Surinamese descent; one grandparent was Chinese. His short black hair is cut in what look like round rosettes all over his head, and he tells me that he cuts his hair himself but permits a client to cut the back. He is helped by a young woman who works as a receptionist and cashier, his nephew sometimes cuts men's hair, and a niece can be called to come in if a client, usually a woman client, wants her hair braided.

Many of Ricardo's clients are professional soccer players. Often they are famous players for teams throughout Europe, the kind of players whose clothes and hair are copied by thousands of young male fans and whose wives and girlfriends are fashion models. Sometimes clients ask him to copy the hair style he has created for one of the pro players whom they have seen on television or in a sports magazine. "I cut this guy's hair four years ago"—Ricardo gestures toward a magazine cover with a professional soccer player on it—"a Mohawk on top with a fade on the sides. People see him on television, and now everyone wants it." Other clients are lawyers or professors who live in the neighborhood, while foreign tourists also find their way to the shop.

Ricardo had a chance to enter this world because he was an amateur soccer player who also worked as a personal trainer at a gym in Amsterdam. His hobby was cutting hair, which he did informally, off the books. He started by helping a friend, and then gained more clients by word of mouth. His boss paid for him to go to barber's school, but Ricardo says he learned everything they could teach in a four-year course in only one year.

Clients came to Ricardo's house for their haircuts, and he continued cutting their hair because he liked to help people and he liked creating styles. Ricardo's big break came when a famous soccer player arrived at his house for a cut. He came because Ricardo had cut his friend's hair, and he liked the style. Ricardo was surprised to see him, but rose to the challenge. "I changed his whole face. He had a bad haircut, and I changed him from a child into a man. Now people want to be like him."

Ricardo had not planned to open a shop. But in 2007, a man came to him and offered to sell him his barber shop at a low price because he wanted to leave Holland. "When I saw the shop," Ricardo says, "I saw the future. I saw my eyes get open." As clients bring him more clients, he plans to expand. "I'll build a business for my family. I'll have Kenza products, Kenza modeling, I'll work for everybody."

Comfort Caffe, Javastraat: Gentrification

Comfort Caffe is a small, Italian restaurant located just off Javastraat. Although the main courses are quite affordable, the imported ingredients from Italy and elaborate wine menu establish Comfort Caffe as an up-market venue. Cream-colored walls, wooden furniture, and a courtyard give the restaurant a coherent, attractive look, which is complemented by a changing exhibition of paintings on the main wall. The thought put into the restaurant's design is emphasized on its website, which offers a 360° view of the restaurant and courtyard, a short film with photo-impressions, and information on the artworks. Reviewers say the small, open kitchen provides a "homey" atmosphere.

The restaurant is run by a native Dutch couple, Chris and Sarah, who lived in the area for years before opening the restaurant. Sarah has a background in theatre, but always combined this with hospitality jobs, mostly "front" jobs as a host or bartender. At some point she felt she had to choose between careers, and decided to open a hospitality business. Her current life and business partner was trained as a chef, although he was working as a mixologist before they opened the restaurant. Their choice to locate in the Indische Buurt was a very conscious one, and was highly influenced by the area's gentrification.

Sarah had initially moved to the Indische Buurt to live with her partner, who already had an apartment there. Coming from a fancier part of town, she felt she was moving into "a ghetto." Despite living there, she and her boyfriend worked and socialized elsewhere. The idea to locate their new business in the Indische Buurt only occurred to them after they had seen considerable changes in the neighborhood.

They were especially interested in the new residents—whom Sarah refers to as a "mix" of residents rather than one particular group. However, because she saw that the new residents were often "Dutch" and younger, first-time apartment buyers, she began to think that the Indische Buurt could become the new Pijp, an offbeat, previously working class and immigrant neighborhood which has been gentrified.

Despite expectations of gentrification, commercial rents were still relatively low in the Indische Buurt. So Sarah, like new retail entrepreneurs elsewhere, saw opening a restaurant as both a good business investment and a chance to "do something for the neighborhood" by contributing to its revival.

After a short investigation into Javastraat, its shops, and customers, Sarah discussed her business plan with the street manager. The manager told her about the empty storefront where she would eventually open Comfort Caffe. Already zoned for a restaurant or café, the storefront had previously housed a Turkish coffee house but was now vacant.

But the building was in such a bad condition that Sarah refused to consider it. At that point, the building owner, a housing association, offered to renovate the building for her. The costs of the renovation are not reflected in the rent, which remains below market level. Sarah also saved money because she took over an empty storefront, and didn't have to pay takeover costs for "goodwill." She also did all of the interior design herself.

Comfort Caffe almost immediately gained a clientele of regular local customers, whom Sarah describes as "young, white families." She even calls them her "family," and, emphasizing the number of young families among them, claims that they show her the ultrasound images of their unborn children before they show them to their own parents.

Like Youssef at Tigris & Eufraat, Sarah has adapted her business strategy to customers' requests. Although the restaurant was originally planned to be a lunch café, focusing on good coffees and a quick bite, local customers started to ask for longer opening hours and a more varied menu. They also posted very positive reviews on Iens, a well-known Dutch website (www.iens.nl), causing Comfort Caffe to end up in the top 10 best reviewed restaurants in Amsterdam. Many of the positive reviews state that the restaurant is an asset to the Indische Buurt.

Since Comfort Caffe entered the Iens top 10, visitors come from other parts of the city and even from affluent villages near Amsterdam. Changing the café into a full-scale, up-market restaurant turned out to be a smart move, because it significantly reduced the number of competitors. In fact, the owners of Comfort Caffe do not network much with other business owners in Javastraat. Neither have they joined the street association, because they regard it to be both "rigid" and insignificant.

Sarah shares the stereotype that there are "too many greengrocers" on Javastraat, and she hopes, as both a business owner and a resident, for a "more varied" selection of shops. When a large electronics store closed about a year ago, she attempted to lobby for the introduction of a bicycle store owned by a friend. However, the rent was too high, and the storefront was taken by a "Turkish" furniture store, which she considers a poor alternative.

This conflict of values between immigrants and gentrifiers also plays out in Berlin.

Acknowledgements

This chapter is based on interviews, ethnographic observations, and a walking census of stores carried out from 2010 to 2014 by Iris Hagemans (Javastraat), and Sharon Zukin and Anke Hendriks (Utrechtsestraat). We express our gratitude and appreciation to all the business owners and managers who generously spent so much time with us, and former street managers Nel de Jager and Marcia van de Hart. We also thank the Urban Studies Program at the University of Amsterdam for its support, and Aart van Duren (Bureau Stedelijke Planning bv), Rogier van der Groep and Peter van Hinte (Research and Statistics Department, Municipality of Amsterdam), and Kamer van Koophandel for help with data collection.

Notes

1 A note on terms: the Dutch use the word *migrants* rather than *immigrants* to classify residents born overseas, which is especially appropriate when it refers to Surinamese, who come to the Netherlands from the former Dutch colony of Surinam and hold Dutch citizenship. Official discourse also avoids explicit racialized terms, preferring to distinguish between people who are *autochtoon* (of Dutch heritage) and *allochtoon* (of foreign birth), although in practice the first category often refers to people whom Americans might call "white" Dutch and the second category often refers to "ethnic minorities" or "people of color" (see Yanow and van der Haar 2013). In everyday language, the Dutch often do use racialized terms such as "black" when referring to immigrant ethnic minorities (Rath 1993, 1999).

2 Numbers and percentages are based on the population statistics of the A03 and A07 districts (i.e. Grachtengordel-Zuid and Weteringschans respectively) on January 1, 2013.

3 Though a street manager may work many hours, both advocating for business owners to various city government agencies and explaining the views of the city government back down to business owners, their salary is not high. So de Jager worked two jobs, as the street manager for both Utrechtsestraat and Haarlemmerstraat.

4 Residents who are displaced are relocated to similar apartments, but they are usually in other, low-rent districts of the city.

References

Amsterdam.info. Undated. "Dappermarkt in Amsterdam." Available at www.amsterdam.info/markets/dappermarkt (accessed November 25, 2014).

Arnoldussen, Paul. 1996. *Het graf van de koopman: Verhalen uit de Utrechtsestraat.* Amsterdam: Jan Mets.

Arnoldussen, Paul. 2011. "Wie kent dit verkeersbord nog?". *Het Parool* (October 17).

Boterman, W. R. and W. P. C. van Gent. 2014. "Housing Liberalization and Gentrification: The Social Effects of Tenure Conversions in Amsterdam." *Tijdschrift voor Economische en Sociale Geografie* 105(2): 140–60.

Bureau Onderzoek en Statistiek. 2013. *Stadsdelen in cijfers 2013.* Amsterdam: Gemeente Amsterdam.

DTZ Zadelhoff. 2013. "Winkelhuren Amsterdam 2013." Available at www.amsterdam.nl/publish/pages/435544/winkelhuren amsterdam_2013.pdf (accessed January 29, 2014).

Ernst, Olaf and Brian Doucet. 2014. "A Window on the (Changing) Neighbourhood: The Role of Pubs in the Contested Spaces of Gentrification." *Tijdschrift voor Economische en Sociale Geografie* 105(2): 189–205.

Funda. Undated. "Homes for Sale." Available at www.funda.nl (accessed January 31, 2014).

Heijdra, Ton. 2000. *Zeeburg. Geschiedenis van de Indische Buurt en het Oostelijk Havengebied.* Alkmaar: Illiano.

Kok, Auke. 2014. "Domweg ongelukkig in de Javastraat." *Vrij Nederland* (January 6). Available at www.vn.nl/Archief/Samenleving/Artikel-Samenleving/Domweg-ongelukkig-in-de-Javastraat-1.htm (accessed May 2, 2014).

Lesger, Clé. 2007. "De locatie van het Amsterdamse winkelbedrijf in de achttiende eeuw." *Tijdschrift voor Sociale en Economische Geschiedenis* 4(4): 35–70.

Parool. 2014. "Bewoners Javastraat kunnen door lawaai terras niet meer in eigen bed slapen." August 24. Available at www.parool.nl/parool/nl/4030/AMSTERDAM-OOST/article/detail/3723958/2014/08/24/Bewoners-Javastraat-kunnen-door-lawaai-terras-niet-meer-in-eigen-bed-slapen.dhtml (accessed September 15, 2014).

Peck, Jamie. 2012. "Recreative City: Amsterdam, Vehicular Ideas and the Adaptive Spaces of Creativity Policy." *International Journal of Urban and Regional Research* 36(3): 462–85.

Rath, Jan. 1993. "The Ideological Representation of Migrant Workers in Europe: A Matter of Racialisation Only?" In *Racism and Migration in Western Europe*, edited by John Wrench and John Solomos, pp. 215–32. Oxford: Berg.

Rath, Jan. 1999. "The Netherlands. A Dutch Treat for Anti-Social Families and Immigrant Ethnic Minorities." In *The European Union and Migrant Labour*, edited by Mike Cole and Gareth Dale, pp. 147–70. Oxford: Berg.

Smit, V. J. M. 1991. *De verdeling van woningen: een kwestie van onderhandelen.* Bouwstenen 21. Eindhoven: Technische Universiteit Eindhoven, Faculteit Bouwkunde.

Staatsblad van het Koninkrijk der Nederlanden. 2009. "165." Available at http://biz.joostmenger.nl/uploads/experimentenwet%20BIZ.pdf (accessed October 22, 2014).

Stadsdeel Zeeburg. 2007. *Convenant vernieuwing Indische Buurt 2007–2010.* Amsterdam: Gemeente Amsterdam.

Van der Pennen, T. and K. Wuertz. 1985. *Bouwen voor de kleurrijke buurt. Allochtonen en besluitvorming in de stadsvernieuwing.* The Hague: VUGA.

Van Duren, Aart. 1995. *De dynamiek van het constant. Over de flexibiliteit van de Amsterdamse binnenstad als economische plaats.* Utrecht: Van Arkel.

Yanow, Dvora and Marleen van der Haar. 2013. "People Out of Place: Allochthony and Autochthony in Netherlands Identity Discourse—Metaphors and Categories in Action." *Journal of International Relations and Development* 16(2): 227–61.

Zukin, Sharon. 2010. *Naked City: The Death and Life of Authentic Urban Places.* New York: Oxford University Press.

Zukin, Sharon. 2012. "The Social Production of Urban Cultural Heritage: Identity and Ecosystem on an Amsterdam Shopping Street." *City, Culture and Society* 3(4): 281–91.

CHAPTER 5

Life and Death of the Great Regeneration Vision

*Diversity, Decay, and Upgrading in Berlin's
Ordinary Shopping Streets*

CHRISTINE HENTSCHEL AND TALJA BLOKLAND

Nicola is the owner of a little café with five gambling machines in a popular Neukölln shopping street. His place advertises itself as open "24-hours non-stop," and indeed, whether day or night, he is there when I walk in. Sometimes he takes a nap, he says, when there are no customers, and during the day his girlfriend comes to relieve him for a bit so he can go home to shower and come back—not a single night of regular sleep in the last eight months for him.

One early morning, I asked him why he opened his gambling café here, in Neukölln. "Because here in Neukölln, there are so many 'socials'," he replies. "Socially weak people?" I ask. "Yes, socially weak people who gamble."

Nicola lives in Lichtenberg, a suburban area in the east of the city, where life is "quiet and orderly." He is from Bulgaria and in his early thirties, and says he would much rather own a cocktail bar than a gambling parlor. He finds the slot machines on his premises "ugly," but they help him sustain an income he could never be able to secure by selling coffee or beer alone. Still, not much remains after paying the rent for the premises and the gambling machines (240 euros for each machine per month to the manufacturers of the machines) as well as a 20 percent tax to the city on all gains.

That night no one gambles and Nicola and a good handful of others celebrate the birthday of a young wrestling champion from Poland. Someone brought take-away food from a nearby kebab place. On the television half-naked women are dancing. A Turkish man who owns a casino in a different neighborhood points out to me how smoothly everybody switches between Turkish, Polish, Bulgarian and Russian.

Indeed, the cosmopolitanism of this crowd was remarkable, different, to be sure, from the cosmopolitanism of the well-educated creative types that northern Neukölln has come to be associated with in the last few years, and from the "diversity" that urban managers trumpet to be Neukölln's asset. What I encounter in this gambling café seems more like a quiet cosmopolitanism at the edge of urban life.

Like Nicola, many small-time entrepreneurs have set up shop in the commercial streets of Berlin's low-income, immigrant neighborhoods. Cell phone shops, dollar stores, bubble tea bars, kebab restaurants, and cheap fashion outlets have transformed the look and the feel of the old West Berlin shopping street, or the idea of it, with its German family-owned "quality" shops and department stores. In this age of high-speed, post-Fordist capitalism, *local* shopping streets are rarely the strongholds of a neighborhood's "authentic" character. Rather, they mirror and foreshadow the effects of globalization—change that does not always fall into the dominant narrative produced by urban studies.

Urban theory on Berlin's transformation operates either with gentrification as a narrative of change or with segregation and ethnic concentration as an unwanted reality. By focusing on such broad homogenizing narratives, these writings miss other equally important, perhaps "messier" processes of transformation, which cannot be boiled down to mere gentrification or segregation. Answering Jenny Robinson's (2006) call to ground urban studies in a greater diversity of urban experiences—not only by studying cities at the world's "periphery" but also by studying the quieter places and processes at the periphery of our dominant reading of the richest global cities—we want to understand the conflicting narratives driving change in two local shopping streets in two disadvantaged neighborhoods of former West Berlin: Karl-Marx-Straße in Neukölln and Müllerstraße in Wedding.

Both streets are, in Robinson's sense, ordinary and unspectacular. They lie at the social periphery of the new Berlin. Those looking for the quaint stores of olden times, like the butcher and the corset shop, where locals mingle and chat, will be disappointed. Gambling and betting parlors, cheap barber shops or one-Euro coffee shops have increasingly become the social space for interaction and passing time.

A narrative of decay structures the discourse on both streets. In Müllerstraße, this narrative encompasses both the street itself and the surrounding area. Karl-Marx-Straße is still seen as in a downward spiral, while the streets surrounding it are dominated by a narrative of gentrification. Diagnosed as difficult areas, yet playing a central role in the district, both streets were assigned the status of *Sanierungsgebiet* (regeneration area) in 2011, which entails an imperative of change as well as funding and new governmental structures to facilitate this change.

The local actors of the regeneration agencies work under the assumption that the support and involvement of shopkeepers are crucial to the future of the shopping street as a lively social space of consumption. But they have had only limited success in getting shopkeepers and residents involved. According to these regeneration officials, who are the street's political agents of change, most shopkeepers seem passive, disinterested, and not keen on shaping the future of their shopping street. We want to know why this is so, and how contrasting interpretations of transformation by shopkeepers and officials interact.

Most local shopkeepers appear disengaged from the revitalization process. They generally do not attend meetings, show interest in developing plans or even in forming a shopkeepers' association. In explaining this disengagement they tend to invoke narratives of fragmented citizenship, exogenous change, and a lost sense of community. Comparing these with the narratives of the regeneration officials, we identify a racist discourse that glorifies the "good old days" of German mom-and-pop stores, that ignores that things have changed, and misses the chance to make the most of current conditions in Karl-Marx-Straße, but may be a blessing disguised as "failure" in Müllerstraße.

Paradoxically, the slim chance that Müllerstraße will ever become truly hip and gentrified creates space for a meaning of "diversity" that seems, as our preliminary analyses suggest, a better fit with the various repertoires of shopkeepers.

We identify two striking misunderstandings between shopkeepers and official agents of change. The first misunderstanding revolves around the meaning of *diversity*: the regeneration officials interpret diversity as making space for more high-end shops and celebrate it on a symbolic level, while neglecting the real social diversity on the ground that needs a different approach and sensitivity. The second misunderstanding concerns the notion of *time*: while renewal officials want to forge the street's *future*, the shopkeepers barely get by in *the present*, or they reminisce about a supposedly better *past*. All of this suggests that working with, not against, the "messiness" of contemporary urban life is important.

Two Streets in Transformation

Located in working-class districts, both streets tell a story of Fordist capitalism, immigration, and post-Fordist decline. Moreover, both streets have seen their roles redefined after the collapse of the Berlin Wall in 1989: first, through a brief flourishing as East Berliners found their way to these nearby and now suddenly accessible shopping streets; and then, through economic decline due to the development of shopping malls on empty land where the Wall had stood.

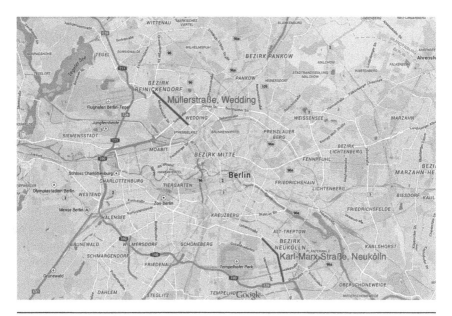

Figure 1   Map of Berlin, Showing Karl-Marx-Straße and Müllerstraße
Source: Google Maps.

Karl-Marx-Straße

Karl-Marx-Straße in Neukölln was built near the turn of the twentieth century when four- and five-floor apartment blocks were erected to house the exploding population of workers from the surrounding factories. From the beginning, it was a social-democrat, communist, and trade union stronghold, hence its naming after Karl Marx in 1947, which is unusual for a street in West Berlin. Many of the buildings had impressive art deco façades, while the interiors were extremely simple. On the ground floors were family-owned shops, for leather products, flowers, musical instruments, photography, and shoes. The lone Jewish-owned department store, H. Joseph & Co., was shut down under Nazi rule in 1936. In 1952, the Hertie chain opened in the former Joseph & Co. building and remained a landmark of low- and middle-income commerce until it went bankrupt in 2006.

In the 1970s and 1980s, Karl-Marx-Straße was the third most important shopping street in West Berlin, retaining a mixture of small-scale family-owned shops and a few department stores as well as public buildings such as the post office, the district town hall, and an opera house. The year 1989 was a grand moment for Karl-Marx-Straße, when freshly freed East Germans thronged to the shopping street. Long-term residents of Karl-Marx-Straße remember long queues in front of the department stores. For a while, there was even a direct bus route from the former East German city of Cottbus, which is 200 km away, to a stop right in front of the Hertie department store. With the opening of the border, business in the street boomed (interview, regeneration manager, Aktion!, Karl-Marx-Straße, 2011).

But the boom killed both the small shops and the department stores. In the early 1990s, the rents for commercial spaces rose drastically. As a result, many small shops couldn't survive any longer and went bankrupt. At the same time, with Germany reunified, many shopping centers opened at the outskirts of the city in former East Germany, just a few minutes from Neukölln, which pulled away a great deal of the local purchasing power. In addition, municipal support for Neukölln was stopped and transferred to urban renewal initiatives in East Berlin.

Local officials tell a story of slow death, reaching the nadir in 2006. The economic and public landmarks in Karl-Marx-Straße closed down, including Hertie and the post office, all of them in the central area of Karl-Marx-Straße. The street was left with secondhand shops, cell-phone dealers, rummage sales, and dollar stores. "This was the absolute anticlimax, the dance of death" (interview, regeneration manager, Aktion!, Karl-Marx-Straße, 2011).

After the "dance of death" came a new kind of life that many still regard with suspicion. Today the street looks as lively and feisty as can be; what died was perhaps more a particular kind of West Berlin shopping boulevard. Before the 1990s, most shops on the street were German-owned, but by 2008, 28.4 percent of the shops were owned by shopkeepers with a migration background—some with German citizenship, some without. Over half of the migrant-owned shops on the street are Turkish, followed by Arab, Polish, Chinese, and Vietnamese (Kayser *et al.* 2008: 34ff.; see Figure 2).

The latest profiling by the district's commercial support unit counted 38 fast food restaurants, 20 fashion stores, 15 bakeries, 14 cell phone shops, 14 pharmacies, 14 call shops, 13 dollar stores, 13 bars, 12 banks, 11 jewelers, and 11 hairdressers (Commercial Support Unit 2006; City Management 2013). Most of them cater to a low-income population. The unemployment rate in the area is around 21 percent (compared with 15 percent in Berlin); 31.7 percent of residents in the direct environment of Karl-Marx-Straße receive state welfare (compared to 14 percent in Berlin). Forty-one percent of the local population has no German citizenship (15 percent more than in 2000). Half of the non-native-born are of Turkish origin; 20 percent come from former Yugoslavia, and 14 percent come from Arab countries (Berlin Senate 2011).

Müllerstraße

Müllerstraße, on the other hand, had been the connecting street between Oranienburg, a small town with a country estate for the Royals outside Berlin, and the city center since 1827. A four-lane street, it has since yielded this

Figure 2   Streetscape, Karl-Marx-Straße
Source: Photo by Christine Hentschel.

function to the various highways around the city, but is still very busy. Leopoldplatz, its central square, functions both as a site for socializing for various groups, including drinkers and homeless people, as well as a point of transit, with a metro stop. With department stores; shoe, clothing and jewelry stores; hardware stores; and restaurants, Müllerstraße was where residents from the entire northern part of West Berlin would shop, and many stores for daily needs were located on the side streets.

This is no longer the case. Now many of the stores on Müllerstraße serve daily needs, and all but one of the big department and clothing stores have closed.

The street had been an important shopping street before the Wall came down in 1989 because there were few others. Two similar streets were located much farther away, and posh Kurfürstendamm, to the west, never served the shopping needs of low-income people. After the Wall came down, though many local residents saw their incomes decline due to deindustrialization, shoppers arrived from former East Berlin. However, in the late 1990s, when shopping areas nearer the city center and the main shopping mall in Gesundbrunnen, where East and West Berlin had been divided, brought more desirable alternatives, Müllerstraße lost out. Yet another mall soon opened in a huge old factory building in Reinickendorf, just to the north of Wedding.

As in Karl-Marx-Straße, the specialty shops vanished from Müllerstraße during the last two decades. Cheap chain stores, kebab restaurants, and casinos have taken their place. As a shopkeeper on Müllerstraße explains (interview, 2011):

> Yeah, Wedding was a real working-class district, and people were very proud to live here. And they were proud of their Müllerstraße, too. That's really changed. There were specialty stores like flower shops or jewelers. Some music stores are actually still left, surprisingly enough. But the specialty stores, for the most part, have closed. Or bakeries, real bakeries. Instead, we now have dollar shops and chain bakeries …
>
> Over there, at Leopoldplatz, right where the grocer is, it's not exactly a favorable advertisement, because it looks pretty trashy. There are no professional displays, in terms of decoration and everything else. Those things do affect the overall streetscape. We also have too many casinos. They're just mushrooming and giving Müllerstraße the wrong appearance.

Müllerstraße, just like Karl-Marx-Straße, is now described using narratives of death. The shopkeeper says:

> The street just continues to die. As you can see, the stores are closing and all of that. More and more junk is coming here; more junk stores are opening up. There used to be more department stores here. Everything was a little more upscale, and the stores were just better. Since then, the street has just kept going down.

Like Karl-Marx-Straße, the metaphor of Müllerstraße's slow death does not mean the loss of residents, shoppers, or shops. Between 2003 and 2008 the population increased by 4.8 percent and is still on the rise. What does look bleak, though, is the socioeconomic situation. The unemployment rate in the street and its direct proximity lies at 14.2 percent, and 21 percent of the local population depends on state welfare (Berlin Senate 2011: 47ff.). But in both streets, regeneration initiatives have set out to bring improvement.

Regeneration Officials and Their Narrative: Neukölln's "Diversity"

Since May 2011, Karl-Marx-Straße has been an area of "urban regeneration," which refers not only to a renewal of infrastructure but to the complete re-conceptualization of the shopping street itself. A three-person team of private urban planners called "city management" has been commissioned by the

Figure 3 Multi-Purpose Store, Karl-Marx-Straße
Source: Photo by Christine Hentschel.

Figure 4 Streetscape, Muellerstraße
Source: Photo by Christine Hentschel.

municipality with developing a strategy to turn Karl-Marx-Straße into a "more attractive" shopping street, where not only locals but also visitors from the rest of Berlin come to "stroll and shop" (interviews, city management and Aktion!, Karl-Marx-Straße, 2011). To develop such a strategy, the city managers have initiated a participatory process in which shopkeepers, residents, and property owners are invited to meet and discuss the street and its future.

This institutional novelty is happening at a time when the residential area around Karl-Marx-Straße, northern Neukölln, is gentrifying. Expats, artists, and students have moved into the area, on the lookout for cheap rents and affordable studio space. Neukölln is the most recent hotspot of gentrification that has been moving through Berlin, a wave that began in Kreuzberg in the 1980s, then moved through Mitte in the early 1990s, affected Prenzlauer Berg in the late 1990s and Friedrichshain in the early 2000s, before it finally reached Neukölln (see Holm 2011). The city magazine *Tip* recently compared Neukölln with New York's Lower East Side in the 1970s, calling the area a playground for the avant-garde (Slaski 2010).

Surrounded by talk of gentrification, Karl-Marx-Straße itself has been spared the arrival of art galleries, clubs, and bars. But the regeneration initiative has created a vision for Karl-Marx-Straße that positions it at the center of its gentrifying surroundings. A free, annually published booklet articulates this vision as *"Broadway Neukölln."* It sounds ironic that the street from which so many shop owners fled in the 1990s and early 2000s, and that has since become a landscape of cheap clothing stores, cell phone shops, and secondhand stores, is now being re-imagined as like the Broadway of New York. But *Broadway Neukölln* is less a description than a vision, with a twinkle in its eye, just like the slogan it comes with: "young, colorful, successful." In a public meeting, the manager of the regeneration team explains: "Young we are, colorful too," but "successful we have yet to become" (panel discussion, zu Hause e.V., 2011).

Success, according to the initiative, lies in "more diversity." But this diversity does not refer to ethnicity or national origins, in a street where 41 percent of residents do not have German citizenship. The "diversity" meant here is a "commercial mix" and the spreading of "quality" into a landscape of low-quality shops.

The city manager sees the opening of the organic shop–bistro "bioase44" in spring 2013 as well as the potential interest of a book store and a vegan restaurant to set up shop in the street as signs of hope (Aktion! 2013: 15). Such shops will "support a healthy commercial mix with the desired individuality" of Neukölln's growing "group of lifestyle-cosmopolitans," as she calls students, singles, and non-married couples with a good education and a high income level, whose consumption revolves around "lifestyle and ambience." Her notion of "healthy" contrasts with the implicit idea that, as it is now, the street may not exactly be dying, but it is severely ill.

Another representative of the renewal initiative agrees: "The goal, of course, is that the dollar stores are gone." Yet, aware of the pitfalls of gentrification, he adds:

> But that would presuppose an increased purchasing power of the residents, so that they can afford the higher prices. It would be nice to magically improve the educational and income situation of the residents in order for them to be able to afford higher quality goods (interview, regeneration initiative employee, Karl-Marx-Straße, 2011).

In order to minimize the number of dollar stores, the regeneration initiative asked a research team to investigate the shopping preferences of potential customers, meaning people who spend their time already shopping or window-shopping in Karl-Marx-Straße. A local official told us that he mobilized the 2,000 employees of the district town hall, which is on Karl-Marx-Straße, to participate in the survey, so that their more middle-class needs would be represented alongside those of low income residents. This would lend credence to the claim that there is demand for more "quality shops."

Going beyond a survey, the initiative is looking for a legal way to prevent more cell phone shops and gambling casinos from opening and bring in higher-quality shops. It is currently developing criteria for the district administration to decide, when a tenant of a storefront moves out or goes bankrupt, who can or cannot replace them.

In addition to this understanding of "diversity" in a purely commercial sense, at the higher end of the market, there is also a symbolic embracing of the term. For example, a plaza on Karl-Marx-Straße has been remade with mosaics in different colors symbolizing the 160 nations living in this part of Neukölln. By appropriating diversity not in social or actually lived ethnic terms but through commercial upgrading and romantic symbolism, the regeneration managers miss the everyday diversity on the street.

Wedding: Lower Expectations

In contrast to Neukölln's ongoing gentrification and, hence, the expected rebirth of Karl-Marx-Straße as Broadway, there is less confidence in Wedding and Müllerstraße. The first students and artists landed in Wedding some years ago, but it is too early to trumpet the turn-around of the neighborhood. Perhaps because expectations are linked to the upgrading of the surrounding residential neighborhood, the politics of change differ from those in Neukölln.

The regeneration plans for Müllerstraße also aim for a more attractive shopping street with a commercial mix, and an upgraded infrastructure. But the planners charged with the regeneration process take a more modest approach to what they can achieve. The perception that Müllerstraße is not, and perhaps will never be, a "glorious boulevard" is widely shared. So the planners want to "work with whoever is there, whether it is a vegetable seller, a café or a chain bakery … These are the structures and we want to reach *everyone*" (interview, shopping street manager, 2013). They see their own role as informers and inter-preters of the processes of demographic and economic change to the shopkeepers, residents, and property owners in Müllerstraße, rather than as actual *makers* of change. As one of the planners succinctly says, "About what we, ourselves, want for the street, nobody gives a damn. You see, I can wish for an organic food store to be here but it is of no use [to wish for that] because we will never be able to steer this anyway" (interview, shopping street manager, 2013).

Since 2008, Müllerstraße's Shopping Street Management has brought together merchants and residents of the street with the aim of building a lobbying group that could represent the commercial interests of the street's stakeholders. After more than four years of monthly meetings, they formally created the group in 2013. It takes time "to agree on what the common interests are," one of the plan-ners explains, pointing to the irresolvable heterogeneity of commercial interests reflected in the debates (interview, regeneration facilitators, 2013).

The meetings we shadowed had a slightly different character and atmos-phere than those in Neukölln. Shopkeepers and facilitators met every month in a different location— a café, a gallery, or a plant nursery, and always at 7:30 am before they opened their shops—while in Karl-Marx-Straße, the shopkeepers met in the offices of the City Managers. In the Müllerstraße meetings, coffee, hand-made cakes, fresh sandwiches or soup were served each time, and the ambiance was always chatty and warm. It was difficult to tell administrative agents of change and shopkeepers apart because the intensity of involvement and concern was widely shared. To be sure, here, too, not enough shopkeepers showed up to make the professionals happy. But there was a more positive embrace of a different kind of diversity, and of the slowness and messiness of the collective process of change.

Shopkeepers' Narratives

"We Need Starbucks"

In some of the shopkeepers' narratives, we see support for the plans and the enthusiasm of the regeneration officials. The owner of a Turkish breakfast restaurant on Karl-Marx-Straße, for example, thinks "there should be more … modern shops, more classic, more romantic [shops]. Not so much crap." She adds, "Yes, we need Starbucks, and coffee shops like that" (interview, 2011). In the café she had opened only eight months before we were talking, she serves fresh pastry, baked potatoes, and fresh juices. She emphasizes that her customers are not only Turks, but lots of Germans too. She is impressed by the success of her own café so far. Until 2010 she had a boutique on the same prem-ises, but, it just did not work. It was too "high-quality, it was just too expensive for this street, for these folks." Starting a new shop was expensive; she went into debt and had to sell an heirloom. But she is sure that it was worth it. Karl-Marx-Straße is "on the rise," slowly.

In Müllerstraße, the owner of a recently opened coffee shop that has another branch in the neighboring, more affluent district of Prenzlauer Berg, sees herself as a pioneer. She expects a better future:

> Everyone has to make their own decision as to whether they trust them-selves and dare to come here. We suffered a whole lot in the first years, when we got here, because everyone said "they don't fit in." And "the only thing that was ever really nice in this building was the optometrist." That is really what, like, the old people would tell us in the beginning. "Who needs you here? … You are not wanted here."
>
> It also depends on age groups. So, the young people then have, some-how, they were happy that we are here. But the elderly, they did not care, because it did not mean anything to them, this kind of coffee house.

Pointing to the local theater, an upscale Italian restaurant that is always busy, and to her own store as well, she says: "So somehow it is nice … You find a few things that you really would not have expected. Like the coffee shop. There are highlights that you would not have expected" (interview, shopkeeper, 2011).

There is also a rare narrative that Müllerstraße is rough and tough, but when polished enough it could turn out to be a diamond:

> Yeah, it's like with pearls or diamonds, when you really polish them. You'd be surprised what Wedding could be. And Müllerstraße, too, if you went about it right. I'm just looking at Müllerstraße and imagining, that, maybe next to the dollar shop, there were a nice organic store or that a deli-catessen had the guts to come here. Or maybe the department store would come back. Or some noble bar would come and say: "Alright, I'll give Wedding a try".

Emiliano is the 60 year-old owner of an Italian delicatessen and he has a longer history of giving Wedding a try. When he opened his shop in the 1990s he felt like a "Porsche car dealer in Albania," and explains:

> The customers were not right for the merchandise I sold. They didn't know what to do with it. I first had to show them what olive oil is, how to eat it how to use it for cooking … Over three years I gave them olive oil and bread to taste (interview, shopkeeper, 2013).

He found his main clientele in a small population of German and French professionals, but with the general crisis in the street in the late 1990s, these customers moved out of the area. Over the next ten years he had to muddle through. Because on Müllerstraße he could not make money with an Italian delicatessen, he turned to outdoor catering for events in more affluent districts.

But things have begun to change for the better on Müllerstraße, he explains. People from already gentrified Prenzlauer Berg have moved to Wedding and frequent his shop. The other day he was surprised to see six young people entering the store and asking for black noodles and walnut pesto. They were from Prenzlauer Berg, but they had learned about the shop through the advertisement brochure *Wedding Weiser,* a guide to local businesses published by the regeneration initiative. Emiliano is positive about the changes happening to his street, and he keeps close relationships with the planners from the regeneration initiative, who often come to his place for coffee.

Such enthusiasm is, however, shared by only a small minority of the shopkeepers that we spoke to. The vast majority showed little interest in the street, had few hopes for its future, and did not want to engage in the professionals' initiatives. These shopkeepers have three different routes to their narratives about not engaging in the street's regeneration.

It's Just a Job

In many cities, a remarkable change since the early twentieth century has been that store owners no longer live behind, above, or even near their stores (Rae 2003). Prior to that time, shopkeepers and assistants did live near their store, and were also active members of the community as churchgoers, parents of schoolchildren, and neighbors. These activities connected them to being engaged in the politics of the street and the wider neighborhood. Political citizenship, in Rae's words, depended on such local embeddedness.

The fragmentation of place of work and place of residence has changed this, and personal and business interests have become geographically separate. The disengaged shopkeepers that we met may be active in their local football club, in their church or mosque, or in the neighborhood association where they live; they may be members of internet support groups, they may save the whales or do something good for society. But their shop is merely a source of income in hard times; it carries little other meaning. As they do not *schmooze* in the local pub, on the church steps, or while waiting at the bakery themselves, they have little more than a business identity in the area. This limits the social meaning of the shopping street.

In both Wedding and Neukölln, quite a few shopkeepers said that they were too busy to go to meetings and did not have much sense of attachment to the area. This was especially true for shop assistants in some of the cheap discount stores that move temporarily into empty storefronts. The high turnover rate of these stores is a business strategy, one that hardly makes local involvement possible.

Others told us that they had few ties to the area because they live elsewhere. They also had no friends or relatives nearby. So they saw their relationship to the street as purely functional, and talked about their store in isolation from the shopping street.

The statement "I am not interested, I do my work and then I leave" (interview, shopkeeper, 2011), may not show an absence of an attachment that can be reinvented. It may be an expression of the role of work in what is an increasingly spatially fragmented urban life.

Deriving no social status from the area itself or from social recognition as members of a local residential community, and without a sense of being part of a community in the shopping street, shopkeepers had no defense against negative stereotypes about the street. This also encouraged dis-identification, a common strategy of persons facing stigmatization (Goffman 1963).

No Sense of Community

The idea of collective efficacy describes the willingness and capabilities of people to work together for the best interests of their neighborhood (Sampson 2012). It refers to mutual trust and the type of community needed to "get things done." But this partly depends on access to information, a topic brought up by some shopkeepers who said they do not care much about what is going on in the street. Moreover, the shopkeepers themselves do not believe they form a cohesive group. They remarked, for example, that "no one participates, there is no interest." They had heard of initiatives, but due to the lack of participation, "Nothing ever came out of that, really."

Moreover, they basically had to learn about such initiatives from the newspaper like everybody else, as there is little communication either among the shopkeepers or between the shopkeepers and the professionals: "You won't see anyone here who does anything. No one passes through here" (interview, shopkeeper, 2011).

Yet others had an understanding that there were good ties between specific shops, but not for the shopping street overall. Their image of community was one of fragmented, multiple circles of support, in which one exchanged small services with the shopkeepers next door, but did not engage with bigger questions regarding the future of the area. They talked about "close contact" with the shopkeepers immediately around them and praised these networks of support: "That is very important, one needs one another time and again," they said. Such contacts were used for getting change, for using a phone when theirs was broken, and other "very simple things."

Nevertheless, there was no sense of community beyond this simple support network. One did not share private worries or business problems, and certainly did not perceive such problems as being of a collective nature. While some shopkeepers reflected on their ties with regular customers as nice, friendly, or even close, and liked the neighboring shops, such ties did not reach beyond their immediate neighbors. Here, too, we see a contrast between economic citizenship and political citizenship. For these shopkeepers, collective action and sociability take place elsewhere.

Some store owners felt there had been somewhat of a community before, and saw their current, neighboring shopkeepers as a threat. On Karl-Marx-Straße, only a handful of family-owned stores of the old West German type remain. The owner of one of these shops, an herbal store more than fifty years old, says she "cannot recognize the street anymore," and that she "cannot go to any café around her shop anymore, because only Turks and Arabs go there and not a single German" (interview, shopkeeper, 2011).

The owner of a Burmese shop, which has been on Karl-Marx-Straße for more than 25 years, misses his walk-in customers. Nowadays the passers-by are Turks and Arabs, and they don't patronize his shop. He doesn't believe change for the better is possible on this street. During our last conversation in early 2012 he feared his neighboring shopkeepers, who came to see him regularly to convince him to give up his shop and even threatened him. By our last visit to the street in late 2012, the Burmese shop had closed down. A shop specializing in muscle-building powders had replaced it.

Likewise, a German butcher on Müllerstraße lamented the changes in Wedding's residential population. Muslims did not patronize his store. He had a plastic pig with a cook's hat in his shop window, creating a clear symbolic boundary of whom his shop was for. He remembers how he used to know everyone, how people recognized him in the street, and how customers used to chat with him more. Though the area may never have been a close-knit community, he experienced it in the past as a comfort zone through public familiarity (Blokland and Nast 2014).

The sense of a lost community is also a basis for disengagement. On Karl-Marx-Straße as on Müllerstraße, negative views of the street and negative experiences on it led shopkeepers to distance themselves. In just a few cases the sense of a lost community has been transformed into engagement. For example, the owner of a tiny German clothing store explains his motivation for attending the business meeting about regeneration by saying: "There are only three German shops left."

To what degree a comfort zone of public familiarity ever existed on Karl-Marx-Straße, we don't know. As an officer in the renewal team points out:

There are people that whine about the loss of these stores, when in fact they often did not contribute to supporting them. They liked the aesthetics, but shopped elsewhere. They just want these stores to exist, because 'they are so nostalgic and homey. It is so nice to have small specialty shops.' Especially if the owner is German and you can imagine having a personal connection to him. But the thing is, most people never had this connection (interview, urban regeneration team employee, 2011).

Whether this personal connection to the shop owners ever actually existed is less relevant than what such stories do: they demonstrate the old timers' disidentification and disengagement. In Müllerstraße, too, many shopkeepers do not think that getting involved in the regeneration initiatives makes much sense. Ensuring the survival of their shop takes most of their time and energy. "Everyone just tries to make their own business work," one shopkeeper noted (interview, 2011).

It Happens Outside

Some shopkeepers explained the lack of solidarity by external processes on which they had no influence. The economic crisis that began in 2008 has increased competition between the stores, and the lower incomes of local shoppers have lowered sales. As the manager of a print shop in Müllerstraße explains (interview, 2011):

> It's a matter of money. People have less money. That's just changed in business. When we still had the deutsche mark, customers had more to spend. You really noticed the change with the euro … You know, back then, unemployment benefits were different. Now, we have Hartz IV [cutbacks in welfare benefits], and we're really noticing that.

The increasingly difficult economic situation has limited shopkeepers' interest in each other and made it even more difficult to think about shared interests: "I think things have gotten worse. Because the prices drop, you see, or the competition is getting worse—or stronger, let me put it that way. So it is more every man for himself" (interview, shopkeeper, Müllerstraße, 2011). These shopkeepers look back to a time when the street had fewer gambling parlors and more German specialty shops. Many express unease about more drinking in public space, drug use and crime. The store manager of a print and cartridge business that has been in Müllerstraße for seven years is positive about his clients and the contacts he has with them, but sees the street as on its way down rather than "rising." He says there is more crime, and the consumption of alcohol on the nearby square has negatively impacted the area. In his view, Müllerstraße has no "flair."

> I don't notice anything that has changed for the better. On the contrary, in the past two or three years we have seen more thefts. That has really increased. Seven years ago we didn't have as many. So we have more thefts. Nothing we can do about that, unfortunately. Other colleagues have the same problem.

I'm just thinking, I know, they've built a playground at Leopoldplatz for the kids, which I think is great, but not too far from that is the drinkers' corner, where many people just get drunk for no reason ... Because they tend to holler and throw bottles, which is a problem, I think ... The people still sit there and get hammered every evening. Not to mention the syringes and drugs they use.

We ask how this affects the shopping street:

Well, many drunken people walk around here. People, who can hardly stand up straight, and then sit down over here in the bakery, not able to tell left from right. Many finance their alcoholism through theft. So, yeah, it definitely has an effect on Müllerstraße.

I would guess, of course I don't know, that you could decrease the number of thefts a bit by approaching the whole problem differently. Because we really have a lot of people here who only walk around drunk. It's not their Friday evening ritual; it's their everyday routine. You can walk past the drinkers' corner any day: they're screaming, throwing stuff. So the problem really hasn't been solved; it's just been relocated (interview, shopkeeper, 2011).

The belief that the economic crisis has reduced the consumption budget of customers, a fact that no upgrading process will change, seems widespread. In Müllerstraße, in particular, this is linked to an understanding of the shopping street as having gone downhill ever since big chain stores started to leave the area. "They would like to have more money to spend," a shopkeeper on Müllerstraße said. Referring to the closing of a branch of a European clothing store chain, he added: "Clearly there was not enough purchasing power out here."

The closure of this chain store was a sign of the downward spiral: "Then C&A shut down, and since I got here, more shops have simply closed. Here in front of us, and there in front of us, and over there is one that's supposedly closing down ..." (interview, shopkeeper, 2011). This affects the overall attractiveness of the street: "It is no longer that interesting here. Especially since C&A has gone. That was a month ago. [Lowering voice] Between the two of us, I would not come here."

Others also referred to the closures when describing how the street has changed. "So Müllerstraße no longer is Müllerstraße [laughs]. It is, with the exception now of Karstadt [another large chain department store] and, eh ... the [chain] drugstore and, and, and our drugstore, besides these, it is just cheap stores."

Shopkeepers, residents, and local bureaucrats all complain about the loss of good stores. This is important because it affects the reputation of the street, how it looks, and what sort of people will come and shop there. Losing "good shops" means no more "good customers." The departure of big chain stores is the last indicator that the place will die (the "dance of death").

Everyone sees a more ethnically diverse neighborhood and street, but the vast majority just express this as a description of change without being negative about it. Some lament the lost kindness of former times; for example, they believe that people and customers were nicer, and people cared more for each other then. Some also see a downward spiral that does not imply a strong notion of a better past, but certainly indicates downward development in the last few years, during which they see increased crime, safety issues, and problems with drinkers and drugs. But they see the downward movement of Müllerstraße as an expression of the overall difficult economic situation.

We see something similar in Neukölln. For those who have settled there recently, like the owner of a Turkish fruit and vegetable store, the street doesn't inspire much hope. When asked about the street, he points to the dollar stores and the gambling parlors around him and concludes that this is not a street to believe in, let alone a place where he should invest his time in community meetings.

Moreover, shops on Karl-Marx-Straße have a very high turnover rate: owners come and go. Many new immigrants have moved into the street. These are no longer just the traditional Turkish and Arab populations, but also Poles, Russians, and former Yugoslavs. To most of them, Neukölln is their first, but not final destination. AbdouMaliq Simone has described how in inner city areas, where people are constantly on the move, the general atmosphere is that of a "state of preparedness" or a "state of suspension"; people are ready to change gears and directions all the time (Simone 2004: 224; 2007: 241). This prevents them from investing in uncertain futures that seem far-fetched and do not lie in their hands.

Surviving in the here-and-now is not just a narrative. It is a strategy dictated by the constraints of urban marginality, especially for those balancing on the edge of legality and economic productivity. Many entrepreneurs in both streets face precariousness on all fronts and engage in a range of semi-legal or illegal activities to make a living while facing a mainstream labor market that provides them with few possibilities, limits their opportunities to open formal businesses, and constrains the purchasing power of their customers. Owners of semi-legal gambling parlors in Neukölln repeatedly explained to us that putting gambling machines in their business is the surest way to make some money. In Müllerstraße, too, for many shops, precariousness is part of the daily business. In a new shop that sells bags and suitcases, the shop assistant, a woman in her fifties, explained that their business basically does "leftover sales." They get small quantities of bags without placing orders for specific products, at irregular times, and try to sell those bags in their small, rather dark store, where the walls are covered by a wide variety of handbags and cases, and larger suitcases are placed in the middle of the floor. As soon as the local market seems saturated, the management will look for another cheap rental space and move on. This is not a business where the shopkeeper will get involved in regeneration plans.

We have identified three routes of shopkeepers' disengagement from local initiatives for regeneration of the shopping street. First, "fragmented citizenship" ends urbanism as it was experienced in the first half of the twentieth century, and makes running a shop a job and nothing more. Second, the lack of a sense of community, outside of a small support network, prevents shopkeepers from developing collective efficacy. Third, shopkeepers explain the downward development of their shopping street through the lens of larger societal developments that, in their eyes, cannot be influenced by local initiatives. Therefore, getting involved in these initiatives makes little sense. We learn from these routes of disengagement that just pointing to shopkeepers' "not caring" is insufficient. We need to understand how they arrive at this position in the context of the street, the city, and modern urban life.

But if streets are intriguing sites of the micro publics of urban life (Hall 2012), scholars as well as urban policy-makers and change makers of all sorts might want to get a better sense of what is at stake in those micro publics. To understand how the street is a place of conviviality and improvisation, a place that encourages change or endures its absence, we need to understand the "messiness" in local shopping streets. Even if we live in, or write about, cities in the Global North like Berlin or New York, we can learn from the informality, chaos, and lack of resources in the Global South (Hentschel 2015). Informality may be at the heart of urbanism everywhere (Roy and AlSayyad 2004).

Curiosity about the "mess" can lead us to better grasp the nature of so-called non-productive businesses like cell phone stores and gambling parlors, and see what they offer to streets and communities. Nicola's gambling parlor in Neukölln pushes us exactly to these kinds of questions. Tapping into a shopping street's assets as a way of reflecting upon, or bringing about, change, requires us to find the owners of the cell phone shops and gambling parlors who do not make it to official community meetings. In other words, an understanding of local shopping streets as "collective" undertakings requires a grasp of the micro-scales of the "intimate street" (Hall 2015: 9). If urban planners and city managers get a sense of what does, or does not, motivate Turkish, Serbian, and German merchants to engage in shaping the future of the street they work in, they can be more humble about the possibilities of change. Perhaps they can then be more open to a broader repertoire of cosmopolitanism. These ideas also apply across the Atlantic Ocean, in Toronto.

Acknowledgements

This chapter is based on interviews with shopkeepers in Karl-Marx-Straße and Müllerstraße, as well as planners, regeneration officials, and so-called city managers active in the transformation of the streets. In Müllerstraße, we worked with students at Humboldt University over two semesters in a project focused on commercial regeneration. The students interviewed shopkeepers about their identification with and participation in the shopping street with special attention to their role as agents of social control, as this was part of a broader project on public safety in the area. Research on Karl-Marx-Straße was part of a larger project on changes in northern Neukölln. In both cases, we also attended neighborhood meetings, spent time talking informally to business owners, and analyzed planning and publicity documents.

References

Aktion! 2013. *Broadway Neukölln 5.* October. Available at http://aktion-kms.de/files/131017_kms-bnk5-131007-rz_lowres_bfua.pdf (accessed March 8, 2015).
Berlin Senate. 2011. Zwölfte Verordnung über die räumliche Festlegung von Sanierungsgebieten. Vorlage des Senats an das Abgeordnetenhaus Berlin. [Twelfth Decree concerning the spatial definition of redevelopment areas. Senate's submission to the Berlin House.] March 15.
Blokland, Talja and Julia Nast. 2014. "From Public Familiarity to Comfort Zone: The Relevance of Absent Ties for Belonging in Mixed Neighbourhoods." *International Journal of Urban and Regional Research* 38(4): 1142–59.
City Management. 2013. *Standortexposé: Karl-Marx-Straße.* Available at https://urldefense.proof-point.com/v2/url?u=http-3A__www.aktion-2Dkms.de_projekte_handel_standortbroschuere_&d=AwIFAw&c=8v77JlHZOYsReeOxyYXDU39VUUzHxyfBUh7fw_ZfBDA&r=IBOqLSbo6B UxgQnX6zQIbbWz1AVlc-0Yclqt_ziogu0&m=7Gqr0EZZbIqbZgnk19nJhcMl9A3oShoupyv USVEnEtw&s=1jrE2t-zWN3uOSUXEtlcr0s59_kp6-ZW5OsjPlP9d2A&e= (accessed March 8, 2015).
Commercial Support Unit. 2006. "Karl-Marx-Straße: Geschäfte." Unpublished document. Neukölln: Commercial Support Unit.
Goffman, Erving. 1963. *Stigma: Notes on the Management of Spoiled Identity.* New York: Simon & Schuster.
Hall, Suzanne. 2012. *City, Street, and Citizen: The Measure of the Ordinary.* London: Routledge.
Hall, Suzanne. 2015. "Super-diverse Street: A 'Trans-Ethnography' Across Migrant Localities." *Ethnic and Racial Studies* 38(1): 22–37.
Hentschel, Christine. 2015. "Postcolonializing Berlin and the Fabrication of the Urban." *International Journal of Urban and Regional Research* 39(1): 79–91.
Holm, Andrej. 2011. "Gentrification in Berlin: Neue Investitionsstrategien und lokale Konflikte." In *Die Besonderheit des Städtischen: Entwicklungen der Stadt(soziologie),* edited by Heike Herrmann, Carsten Keller, Rainer Neef, and Renate Ruhne, 213–32. Wiesbaden: VS Verlag.
Kayser, Peter, Preusse, Faya and Jörg Riedel. 2008. *Ethnische Ökonomie in Neukölln: Zusammenfassende Untersuchung der Erhebungen und Projektergebnisse im Bezirk Neukölln zum Thema ethnische Ökonomie.* Institute for Innovation, Communication and Organisation, Berlin.
Rae, Douglas. 2003. *City: Urbanism and its Ends.* New Haven CT: Yale University Press.
Robinson, Jennifer. 2006. *Ordinary Cities: Between Modernity and Development.* London: Routledge.
Roy, Ananya and Nezar AlSayyad, eds. 2004. *Urban Informality: Transnational Perspectives from the Middle East, Latin America, and South Asia.* Lanham MD: Lexington Books.
Sampson, Robert. 2012. *Great American City. Chicago and the Enduring Neighborhood Effect.* Chicago IL: University of Chicago Press.
Simone, AbdouMaliq. 2004. "People as infrastructure: Intersecting fragments in Johannesburg." *Public Culture* 16(3): 407–29.
Simone, AbdouMaliq. 2007. "Deep in the night the city calls as the Blacks come home to roost." *Theory, Culture, and Society* 24(7–8): 235–48.
Slaski, Jacek 2010. "Spielplatz der Avantgarde." *TIP Berlin* 39(March 4): 29–35.

CHAPTER 6

Toronto's Changing Neighborhoods
Gentrification of Shopping Streets

KATHARINE N. RANKIN, KUNI KAMIZAKI,
AND HEATHER MCLEAN

Like the other cities described in this book, Toronto has experienced major structural changes in the retail sector since the early 1990s. Large-format retailing has filled suburban industrial lands vacated by processes of economic globalization. Big-box stores achieve large "footprints" at relatively low rents, which, combined with innovations in logistics, inventory control and category management, gives them an advantage over small, individually owned shops. Operating with high volumes, narrow margins and low price points, "retail suburbanization" poses a challenge for traditional local shopping streets. Recently, clusters of big-box retailers have formed "power centers" or "power nodes" that further concentrate retail, while big-box formats have also begun to penetrate downtown markets.

Facing continued competition from malls and superstores, local shopping streets suffer from increasing vacancies, as well as a decline in certain types of shops, especially hardware, office products, and general merchandise. Yet they show an increase in food stores, and personal and business services. This selective growth suggests that we are not seeing the "death of the commercial street" in Toronto. In fact, Ryerson University's Centre for the Study of Commercial Areas identified 200 commercial strips with 18,000 shops in 2000 (Jones and Doucet 2000), and a 10 percent increase in the floor area of traditional retail since 2001 (Simmons 2012).

Chains are of course evident, and a few neighborhood shopping strips have been colonized by multinational retail capital. But for the most part, small, independent "mom and pop" stores still prevail. In everyday parlance Toronto is often glossed as a "city of neighborhoods"—and commercial strips are one of the key enabling conditions for the prominence of neighborhoods in everyday life. They provide a range of consumer products and services, public amenities and social services that allow residents to fulfill everyday needs close to home.

Toronto is also recognized as one of the most diverse cities in the world as a result of a half century of immigration from Europe, the Caribbean, Asia, Africa, and Central and South America. As transnational spaces of consumption and exchange, commercial strips play a key role in maintaining the cultural identity and socio-economic stability of neighborhoods. They also furnish opportunities for immigrants to incubate businesses that cater to immigrants' needs, such as for internet hotspots. Some commercial streets in Toronto have created thematic street signs and street festivals that mark and celebrate the area's ethnic identity (Hackworth and Reikers 2005).

Several factors account for the continued vitality of local shopping streets in Toronto. First, the municipal government supports them. When suburban shopping malls began to threaten sales in the 1980s, shopkeepers in a west-end downtown neighborhood formed a business improvement area (BIA) that quickly got the City's support. The founding idea of the BIA, a model that inspired BIDs in the U.S. and elsewhere, was to replicate the malls' strategy for managing shared commercial space. Each BIA collected a voluntary levy from local shopkeepers to fund minor street improvements, and in what turned out to be a significant historical move, they approached the city for matching funding.

This effort developed into a citywide policy to spread the BIA model as "a self-help program aimed at stimulating business" (City of Toronto 2013). Today a mandatory levy on commercial properties furnishes a budget for neighborhood business leaders to expend on local improvements, with opportunities for technical and financial support from the City. Although most BIA annual budgets remain fairly small (less than $170,000), in some cases they have furnished significant opportunities for small, independent businesses to collaborate in local development. They also establish direct lines of communication and accountability with various City departments engaged in planning local streets (Rankin and Delaney 2011).

The second set of factors supporting the vitality of local shopping streets relates to changing demographics and resulting "pockets of resistance" toward retail suburbanization (Simmons 2012). On the one hand, a tripling of the population and more than doubling of personal income since the 1960s, combined with increased household mobility promoted by car-oriented development, has increased demand for more stores consistent with the ambitions of big-box retail. On the other hand, some demographic segments seek out traditional retail and service outlets on neighborhood streets—namely downtown, middle class gentrifiers who buy and renovate old houses, downtown condominium residents, and immigrant populations settling primarily in the more affordable inner suburbs. Each of these populations is growing, which suggests a growth dynamic as well for local shopping streets.

But the spatial distribution of these demographics is sharply divided. A recent study revealed that Toronto is increasingly polarized both demographically and socio-economically into "three cities" (Hulchanski 2010; see Figure 1).

The downtown core, made up of neighborhoods accessible to the city's limited subway network, forms City #1; it includes households with relatively high and increasing incomes and educational status. City #3 forms an inner-suburban ring of poverty, declining incomes, and concentration of immigrant, visible minority and less educated populations. And City #2, located geographically between the others, is the shrinking zone of relatively stable, middle-income households; it is projected to shrink as the other two expand. These social and spatial inequalities reflect processes of economic globalization and restructuring of domestic labor markets shared by the other American and European cities in this book, with many neighborhoods bearing the burden of low-wage jobs that cannot support middle-class consumption.

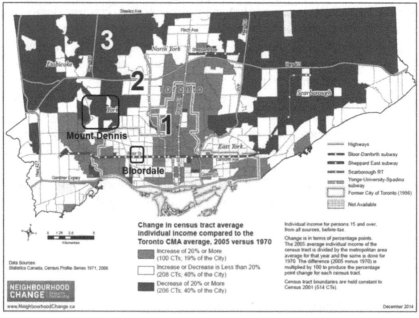

Figure 1 Map of Toronto's "Three Cities," Showing Locations of Bloordale and Mount Dennis

Source: Hulchanski (2010). Reproduced with permission.

Increasing polarization reflects in part mid-twentieth-century planning decisions to concentrate population growth in high-rise towers in Toronto's inner suburbs (City #3), making Toronto home to the second-largest concentration of high-rise residential buildings in North America. Meanwhile, downtown residential property values (in City #1) escalated with demand from high-wage professionals working in the command centers of the service economy—leaving the suburban towers as a more affordable alternative.

Shifts in immigration policy have also intensified spatial inequality. Since the 1960s, the primary source countries for Canadian immigrants shifted from Europe to Asia, Africa, the Caribbean, and Central and South America (Lo 2009). Racism in labor markets, combined with failure of Canadian employers to recognize foreign credentials, has resulted in a concentration of new immigrants in low-income, often off-the-books service jobs. These populations predominate on shopping streets in low-rent neighborhoods like Bloordale and Mount Dennis.[1]

Two Shopping Streets: Bloordale and Mount Dennis

Of our two streets, Bloordale is the downtown strip. It is located on the west end of Bloor Street—the major east–west thoroughfare that cuts through downtown Toronto—under which a major subway line runs. It is one of the few remaining affordable downtown neighborhoods, having only recently been colonized by the ABCs of gentrification, art galleries, vintage fashion boutiques and cafes, as well as the middle-income gentrifiers who typically follow. It may follow a path like that of Orchard Street in New York or Karl-Marx-Straße in Berlin.

Mount Dennis is the inner suburban strip. It falls within the poorest provincial electoral district in Ontario and is known as a neighborhood of immigrant settlement. At the same time, however, Mount Dennis faces considerable pressures for redevelopment, around which various stakeholders are mobilizing competing visions for the future of the neighborhood. In that sense, like Fulton Street, Javastraat, and Muellerstraße in this book, Mount Dennis may face pressures for gentrification.

Bloordale and Mount Dennis share several important characteristics, which reflect how class dynamics intersect the logics and practices of structural racism in the production of commercial space. Both streets have faced the challenge of gradual, long-term, post-industrial disinvestment and have attracted sizable, low-income, new-immigrant populations. With disinvestment, however, came low rents, and merchants and vendors who provide affordable goods and services for new immigrant populations. Unlike middle-class households, who procure goods and services privately, low-income households depend on local shopping streets for a wide range of personal services (Mazer and Rankin 2011; Rankin and McLean forthcoming).

The shopping strips of Bloordale and Mount Dennis also serve as incubators for the low-margin, independent businesses commonly operated by new immigrant entrepreneurs. At the same time, both streets are undergoing significant change: overt gentrification in the case of Bloordale, and revitalization pressures associated with planned transit infrastructure and redevelopment of adjacent vacant industrial land in the case of Mount Dennis.

These challenges to the cultural ecosystem of affordable shopping streets make it urgent to consider how to deflect or mitigate displacement of stores. They also highlight a key difference between our two shopping streets having to do with the dynamics of "criminal insecurity" (Wacquant 2007).

While the law-and-order campaigns that formerly targeted prostitution and drug dealing on Bloordale have dissipated with the creep of gentrification and its associated "social mix," the shopping street in Mount Dennis has become one of Toronto's most intensive targets of police control in the form of "carding." Police officers stop and question passersby in targeted patrol zones, and file document cards with the information they collect. The police justify carding as a strategy to fight drug addiction and youth violence, but we believe it is used to draw attention to "anti-social" individuals in these areas, rather than to reverse the social insecurity there due to labor market deregulation and structural racism, and government welfare retrenchment. These factors are crucial to understand the stigmatization of local shopping streets oriented toward the needs of low-income, racialized immigrants (McKittrick and Wood 2007).

In Canada the discourse of "multiculturalism" has tended to quell scholarly and public deliberation over the mutual imbrications of poverty and race (Galabuzzi 2007; Viswanathan 2010; Roberts and Mahtani 2010). Regarding commercial space in particular, urban geographers have shown how celebrations of "diversity" take the form of neighborhood branding that seeks to commodify ethno-cultural difference (e.g., Goonewardena and Kipfer 2005). Such practices are apparent in gentrifying downtown Bloordale, but in inner-suburban Mount Dennis, structural racism is more evident than ethnic branding.

The selection of a downtown and an inner-suburban street also points to the different policy regimes that govern commercial streets in Toronto. In Bloordale, recent commercial transformation has been largely proceeding in the absence of public policy interventions. Rather, change in the shopping strip has been fuelled by BIA branding strategies and market pressures deriving from the proximity to downtown. By contrast, Mount Dennis was identified in 2005 as one of 13 (now 31) priority neighborhoods targeted for a municipally managed poverty reduction strategy. These differences remind us of the market-driven gentrification of Utrechtsestraat and state-driven gentrification of Javastraat, in Amsterdam, and of similar contrasts between Williamsburg and Harlem, in New York (see Zukin et al. 2009). They also call attention to the role of policy and planning interventions, in general, in shaping local shopping streets (Hall 2012).

Introducing Bloordale

Despite its prime location near the downtown and its excellent access to public transit and other public amenities, Bloordale's image has long been associated with drugs, crime, prostitution and gritty, run-down storefronts. The area is characterized by a concentration of low-income populations, including the "working poor," homeless, and people with mental health and addiction issues. It is also an ethnically diverse neighborhood, where previous concentrations of Portuguese and Italian families have been joined by Indian, Chinese, Burmese, Vietnamese, and Latin American households, many of whom are new immigrants attracted by affordable rents. These low-income and newcomer populations are supported by several social service and settlement agencies in the area.

In contrast with its image as a disinvested area, however, Bloordale has recently undergone a rapid transformation. Since the early 2000s, the neighborhood has experienced a startling rise in property values, and an influx of white, middle-class families, young couples and students. New art galleries, trendy bars and restaurants, and vintage clothing shops have opened. Even the *New York Times*, ever vigilant for trendy urban destinations, has identified Bloordale as one of Toronto's "up-and-coming" districts that would "not go unnoticed by the city's growing creative class" (Kaminer 2012).

History: From Basic Businesses to "Art + Design"

In the 1950s and 1960s, Bloordale was a thriving working-class neighborhood, mainly occupied by Italian and Portuguese immigrants, many of whom walked to work at industrial warehouses and factories in the neighborhood (McBride 2008). The proximity of residence and employment contributed to a small factory town feel. In those early years, the commercial strip was a hub that provided convenient access to basic goods and services for daily needs. Almost half of all businesses were clothing shops, stores selling household goods, laundromats, and dry cleaners. Over 60 percent of food stores were bakeries, delicatessens, butchers and candy stores.

These small retail businesses gradually declined after a horserace track was replaced by a retail plaza in 1956, and this was converted to an enclosed, inner-city mall in the 1970s (Marshall 2013). As many interviewees recall, the opening of Dufferin Mall with its big-box retailers reduced foot traffic on the Bloordale strip. Retail services decreased from 70 establishments in 1960 to 52 in 1970. Specialty food shops were halved, from 16 in 1960 to 8 in 1970.

Another wave of commercial change came in the late 1980s and early 1990s, when nearby factories and warehouses gradually started to close or move out of the neighborhood. The number of Bloordale residents working in manufacturing jobs drastically declined by almost 45 percent from 10,040 in 1981 to 5,735 in 1996, while the area's unemployment rate soared from 5.2 to 13.4

percent. At the same time, the total number of business establishments in the area shrank by almost 20 percent from 1990 to 2000, although restaurants, bars, and stores selling used or discount merchandise increased. Some interviewees believe the disappearance of industrial jobs was strongly related to an increase in trafficking, prostitution, and criminal activities in the area. A number of neighborhood restaurants and bars were associated with criminal activities and routinely targeted for heavy policing.

Until recent years, there were no significant commercial "anchors" on this strip, other than two strip clubs and a large thrift store (Rankin 2008). Most businesses are small, independent stores serving basic needs—dollar or variety stores, pawn shops, shops selling used appliances, check-cashing services, and travel agents. There are some ethnic-identified businesses catering to South Asian and Portuguese communities.

The third wave of change arrived in the mid-2000s as West Queen West, the city-designated "Art + Design District," experienced significant gentrification associated with condo-loft conversions. Art galleries there looked to Bloordale as the next affordable space (Murray 2008).

In 2008, two art galleries moved from West Queen West to Bloordale as self-declared "pioneers." By 2012, four art galleries were concentrated at the west end of the commercial strip, while three additional galleries had moved near the shopping street, into old industrial warehouses that could be converted to large, open spaces. Galleries were soon followed by other "pioneer" businesses such as hipster bars and vintage shops that took the financial risk of establishing in a neighborhood still perceived as "gritty." The entry of the ABCs inspired the 2012 edition of the guidebook *Lonely Planet* to describe Bloordale as "Toronto's coolest neighbourhood" (Mutic 2012).

Introducing Mount Dennis

Mount Dennis, an inner-suburban neighborhood located northwest of downtown, only became part of the City of Toronto in 1998 in a territorial amalgamation. Originally a manufacturing hub and thriving, British, working class community, Mount Dennis is now a disinvested area, struggling with high storefront turnover and vacancy rates. Though rents are low, stores have a low average sales volume, and residents have low incomes and are underemployed.

With 57 percent of the population foreign-born, Mount Dennis is known as an immigrant "landing" area. Well-worn paths from East Africa, Southeast Asia, the Caribbean and Latin America lead to notoriously underserviced highrise apartment buildings and illegal storefront conversions from commercial to residential use. The incidence of violent crime is high relative to the city average, as is sensationalized reporting in the media. Moreover, in 2005, Toronto Police 12 division initiated the practice of carding which is designed to generate a massive database on area youth for "investigative purposes" (Mutic 2012).

History: From Factory Town to Cultural Hub?

Like much of metropolitan Toronto, land in the Mount Dennis area was surrendered to the British Crown by the Mississauga of the New Credit First Nation in 1806, without a proper treaty. But the first immigrant settlement of Mount Dennis followed the establishment of a Kodak manufacturing plant in early 1917. This development catalyzed the growth of an "unplanned blue-collar suburb" made up of both self-built and Kodak-built worker housing and a nearby shopping street catering to the needs of these residents (Harris 1999). Historical sources and key informant interviews attest to the many services Kodak provided to community residents during this era, and also to the vitality of the shopping street, especially because of the lunch-hour traffic that guaranteed patronage of local retail shops and food services (Mount Dennis Community Association 2007).

Kodak employed 3,500 people for nearly 90 years before downsizing dramatically to 320 employees in the early 1990s, then closing completely in 2006. Left behind were 53 acres of vacant "employment land." Even by the 1960s, small independent businesses in Mount Dennis had started to suffer from the emergence of shopping malls and discount stores nearby; the construction of a nearby parkway in 1982 resulted in the further withdrawal of foot traffic from the commercial strip and increases in vacant storefronts (McGinnis 2001). Between 1960 and 1980, the number of clothing and personal goods stores was halved, while the number of specialty food stores was reduced from ten to only two.

The speed of neighborhood decline was further intensified during the 1990s as the impacts of deindustrialization were felt. The gap between average individual income in Mount Dennis and the city as a whole had been increasing since 1981, which was also, not coincidentally, the same period when the number of manufacturing jobs held by Mount Dennis residents started to fall significantly. Nevertheless, this period also marked a major increase in new immigrant populations.

Today, in addition to sensationalizing crime and poverty, the media typically characterize Mount Dennis as "an industrial wasteland" (Monsebraaten 2009) and "the rust belt of the Greater Toronto Area" (Toronto Life undated). The high incidence of poverty has earned Mount Dennis the status of a Priority Neighborhood targeted by the City of Toronto and United Way of Greater Toronto in a place-based approach to poverty alleviation involving all three levels of the Canadian state and numerous nongovernmental agencies.

At the same time, Mount Dennis is attracting considerable pressure for redevelopment. The City of Toronto, real estate developers, and community groups are promoting various visions for boosting the lagging local economy. All of these aim to redevelop the adjacent Kodak lands, and focus on creating a "mobility hub" associated with a planned Light Rail Transit expansion along a main west–east arterial road cutting through the neighborhood.

The visions for redevelopment can be loosely divided between "real estate" and "green-cultural economy" strategies (Rankin, Kamizaki and McLean 2013). "Real estate" advocates, an assemblage of local city councilors, real estate and development industry experts, residents' associations and area BIAs, support residential and commercial intensification. Their recent proposals include a high-density residential development, and new retail and service shops surrounding the planned transit stations—accompanied by intensive place-marketing and branding—as well as a major big-box retail and office park on the former Kodak lands. All of these rest on the premise that the existing population in the area is inadequate, in terms of numbers and purchasing power, to support a viable shopping street. This vision is also built on the notion that if a more affluent population is attracted to the area, better businesses will follow.

The "green-cultural economy" vision also advocates intensive development, and proponents include some of the same actors (residents' associations, BIAs), but also some left-leaning community development advocates and local politicians with deep activist roots in the neighborhood. The "green" dimension derives from a history of environmental activism in Mount Dennis, and seeks to enhance connectivity to the area's greenways, strips of undeveloped set aside for recreational use and environmental protection, that have been historically cut off from the neighborhood. Moreover, supporters have developed an alliance with Blue Green Canada to advocate the promotion of "green jobs" providing living wages on the Kodak lands, and formed a citywide partnership to secure "community benefits" from large-scale public infrastructure development projects.

The vision's "cultural" dimension refers to an initiative to orient transit-leveraged redevelopments toward establishing Mount Dennis as a "cultural hub," which would presumably lure firms in creative sectors from the downtown to the more affordable inner suburbs. Proponents retained Artscape, Canada's leading developer of tenant space for the arts and culture sector, to conduct a feasibility study which advocates "putting challenged neighborhoods on the map for creative people" (Artscape 2011: 41).

Close Up: Bloordale

Changes in the Retail Landscape, 1960–2010

During the past half-century, the number of retail businesses in Bloordale decreased by more than one-third, but the types of businesses remained diverse. In line with broad changes in Toronto's economy, manufacturing businesses in Bloordale practically disappeared, and retail services declined by half, yet personal services grew (see Figure 2). However, the findings of the field survey that we conducted in 2012 suggest that half of the businesses in Bloordale are new, operating for less than 5 years, and 60 percent of them are

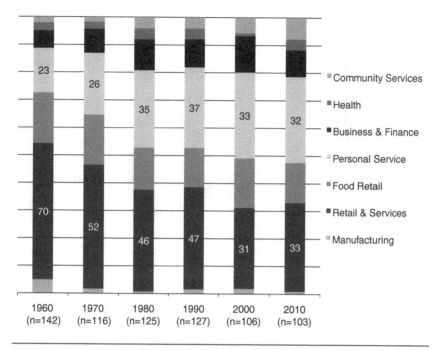

Figure 2 Changes in Retail Businesses, Bloordale BIA, 1960–2010

Source: Might Directories (1960, 1970, 1980, 1989); Polk & Co. (2000); Pitney Bowes
  Mapinfo (2010).

owned by white business owners who tend to cater to more affluent and/or new populations (see Figure 3).

Residential upscaling has taken place simultaneously with retail upscaling. Bloordale and its surroundings have been widely portrayed in the mainstream media as an affordable area for young first-time homeowners (Ireland 2010). In 2012, around 20 planning applications for residential redevelopment, mostly mid-to-high-rise, were under review or approved by the Planning Department (City of Toronto 2012). Some of them are located on the area's industrial lands, which are now being built up with "live-work" residences for gentrifiers (Rankin 2008).

Criminal Insecurity and Social Mix

All of the long-time shopkeepers we spoke to acknowledge the neighborhood's troubled history with crime, prostitution and drug trafficking and identify them as a key challenge for their business. The Bloordale BIA and neighborhood residents' association have developed a strong relationship with the local police division, which significantly increased patrols and arrests during the

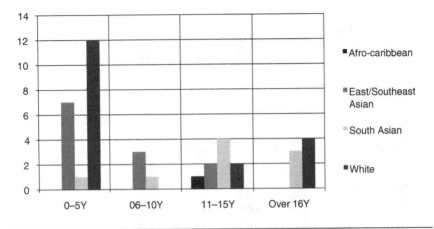

Figure 3 Business Owners' Ethno-Racial Background and Years of Operation in Bloordale
Source: Field surveys, 2012.

early 2000s. This increased surveillance was universally supported by local shopkeepers, who routinely praised the police for their role in reducing crime in the area.

On their own, however, some shopkeepers had developed other strategies for warding off the perceived threat from street-dependent populations. These ranged from handing out coffee and snacks to placing flower pots and other physical obstacles on the sidewalk to prevent street-dependent people from loitering.

Bloordale and its immediate surroundings host several homeless shelters and immigrant support services. Representatives of social service agencies with whom we spoke emphasized the role they play in providing counseling and other services to the members of marginalized and vulnerable groups that are a familiar presence on the shopping street. Remarkably, however, the work of these agencies in keeping social order on the street rarely came up in our interviews with business owners; most were either ignorant of them, or not interested, with the exception of one gallery owner.

Meanwhile, the BIA in Bloordale takes a different approach to criminal insecurity. It is an unusually active BIA, staging a major annual street festival and coordinating a streetscape renovation plan, and it is energized by an unlikely partnership with a handful of resident artists.

The BIA is chaired by the owner of a strip club, one of the oldest, and certainly the most lucrative businesses on the strip. But the inspiration for its major activities comes from a local artist who recognized an opportunity to leverage community art through the BIA mechanism. She persuaded the BIA chair against installing security cameras, shifting instead to sponsor a series of well publicized community art events. These have included exhibits in the windows of vacant storefronts, graffiti removal and painting of façades, partnerships with schools to produce murals and restore ceramic tile on storefronts, and a light-art exhibit featuring citywide artists in the strip club itself, along with more conventional street improvements such as banners featuring local artists.

To be sure, the BIA has continued to work closely with police on targeted surveillance. But the emphasis on community art was intended to address safety by fostering greater community integration, drawing residents and businesses together in small improvement projects that "create destinations for a walk" (Rankin and Delaney 2011). These objectives align with the usual mandate of BIAs, insofar as they also bring publicity to the neighborhood, create a "beachhead" of cultural activity that draws artists into the neighborhood, and ultimately, as the BIA chair consistently emphasizes and publicly celebrates, raise property values.

This strategy of "rebranding" the commercial space has succeeded on all counts. The galleries came, and boutiques and trendy restaurants followed. In interviews with both neighborhood leaders and shopkeepers, the classic "frontier" myth of gentrification surfaced numerous times. Our interviewees spoke of boutiques moving into an "undiscovered area," a trendy restaurant encouraging clientele to "come explore," and consumers' desire to "go hunting" in an area that is "not yet touched" (see Figure 4).

Figure 4 Art Gallery (at the Right) that Moved to Bloordale from Queen Street West
Source: Photo by Brendon Goodmurphy.

These comments are echoed in recent blog coverage of the area; for example a Centennial College student magazine (Gupta 2009) quotes a local store manager as claiming that having art galleries "makes the scumminess of the area seem chic." The new galleries, boutiques and restaurants have been amply reviewed in newspapers' Style and Food sections and have a strong presence on social media and cultural tourism websites like TripAdvisor and Chowhound. All stress the transformation from grit to cultural distinction, and celebrate the courage and character of adventurous entrepreneurs (cf. Zukin 2010).

Given the recent local history with crime, prostitution, and drugs, a majority of business owners whom we interviewed—including immigrant owners who may face displacement—regard neighborhood upscaling as an improvement that has contributed to the decline in crime and violence on the street. The only consistent criticism of gentrification has been voiced by the first art gallery to move into the neighborhood from West Queen West. Amidst the hype of neighborhood transformation, this gallery hosted a very well attended exhibit and town hall discussion on "Demystifying the Creative City," which featured research and analysis seeking to expose the dynamics of occupation and displacement that accompany gentrification (McLean 2009: 208).

In an interview, the gallery's co-owner drew attention to the social ties among long-time shopkeepers, residents, and street-dependent populations. She referred to these ties as an informal "connectedness" allowing for a peaceful and even supportive co-existence. She notes that the affordability of basic goods and services has been a key stabilizing force in a neighborhood wracked by economic decline. And she laments displacement pressures, such as the loud band music coming from one of the area's new hip bar/restaurants, which force the long-term, low-rent residential tenants upstairs to consider moving out.

Mammalian Diving Reflex, a performance art group, has similarly focused on Bloordale and other disinvested West-end neighborhoods, to confront the politics of gentrification by giving voice to stigmatized groups, such as the strippers, drug users and homeless shelter clients, while also revealing the force of displacement pressure in poignant, human terms.

A Newcomer's Narrative: "We're Not Gentrifiers"

Tensions brought on by impending gentrification are the subject of everyday discussion among shopkeepers as well as heated exchanges online. A *Globe and Mail* newspaper article (Hershberg 2011) covering the sale of a long-time diner to gentrifiers, for example, generated numerous critiques of gentrification on local blogs. Referring to this incident, a co-owner of the diner told us that:

> There were some blogs [that] were talking about the new space going in, and they were saying, 'Oh did you see what they did to the [store]front [which the landlord had renovated with matching funds from the BIA]?

Gentrification! They look like Wonder Bread, they're just cookie cutter.' And there seemed to be a very bad vibe and stereotype that we were coming in and were gonna steamroll honest, good working people and the general working class, which is what we are—that we were gonna be a Starbucks.

There is always that word 'gentrification' that seems to be a stereotype in a way and we have been hit with that word, and people trying to knock us down because they haven't come in here they just see brand new shiny and they think that we're a corporation coming in (interview, August 16, 2011).

This interviewee was at pains to clarify that he and his co-owner have kept prices down; they live and work in the neighborhood, and work at three jobs to pay the rent. They do not have rich parents bankrolling the restaurant, but instead spent the money they received as wedding gifts to open it. They are not trying to kick out poor people but trying to be a meeting place between older residents and newer residents: "We are a place that the general public can come to, even just for a cup of coffee."

All of the businesses that might be categorized as "gentrifying" convey a desire to keep the existing social mix. At least they express an interest in "gentrifying differently," much like the "social preservationists" described by Japonica Brown-Saracino (2009). They emphasize maintaining affordable prices, keeping a clean but not a chic appearance, and avoiding the fate of West Queen West and other West end neighborhoods which have recently found a place on the trendy nightlife circuit.

The owner of an organic food boutique spoke with us from a somewhat defensive stance about pressure they had experienced in the neighborhood not to make their store look "too nice":

I guess they [left-leaning critics of gentrification] don't want people to be too successful and they don't want their neighborhood to change too much, just a little bit. But not too much because they don't want their rents to go up, which is totally understandable. And they don't want— they used the word—Yuppies coming in (interview, August 24, 2011).

Defensiveness was coupled with a certain pride in the active role new businesses see themselves playing in redefining neighborhood identity, improving neighborhood safety, and catering to local "community" needs that had not been met by long-time businesses. Some business owners used the idea of "social mix" to defend commercial gentrification. The risk of displacement is rarely mentioned, or, if acknowledged, it is attributed primarily to the logic of "market inevitability."

With gentrification, one of the problems is low-income people are pushed out of the area sometimes, right? … That's a shame if that happens … [But] you know, it happens, right? It is inevitable if prices of the neighborhood [go] up, and some people can't live here anymore … As long as the neighborhood caters to everybody, that's the most important thing, and right now [the Bloordale commercial strip] does … [Displacement] is kind of an economic thing … You can't really stop it from happening once the train is out of the station (interview, August 12, 2011).

Although support for a law-and-order approach to crime and street-dependency clearly prevails among shopkeepers in Bloordale, the art-based orientation of the BIA, the anxiety about gentrification, and the apparent willingness to "pitch" commercial services to a genuine social mix all make it hard to see this as a typical story of commercial gentrification. On this street, at least, awareness of a potential for displacement opens the door to questioning market-driven processes of commercial change. Yet there is no real program of social equity that would withstand pressures for redevelopment and displacement.

Shopkeepers' Stories

Downtown Express, Bloordale: For Both Immigrants and Gentrifiers

This convenience store on the Bloordale commercial strip opened 14 years ago. Since then, the Bangladeshi owners—a married couple and their son, have worked seven days a week, from morning to night, to keep the shop running. The heavy workload leaves no time for leisure. Even though he would like to have more time off, owner Akash does not dream of making more money. "I just want to make a living so I can sleep and live peacefully" he joked.

When the topic of neighborhood change came up, Akash claimed that the gentrification happening in Bloordale was a "good thing." For him, the prostitutes and drug dealers that hung out on his block made his customers and other residents nervous. He explained, "sometimes bad customers come and my wife and I have to handle them. Some people see niceness as a weakness, so you have to be tough with some customers."

On the other hand, he expressed a commitment to helping newcomers trying to make their way in Canada. He took pride in the fact that Downtown Express was a meeting space for the large number of Burmese people who live in the area. He mentioned that Bangladeshi residents built a sense of community by shopping in the store. He was pleased to be part of this community building work, which he considers particularly important for newcomers with limited language skills trying to navigate a new country, especially refugees who might not have had a formal education. In this sense, his approach to running a convenience store mirrored his role as the elected leader of the Bangladeshi expatriate society in Toronto.

Akash also demonstrated an excitement for life-long learning and sharing ideas. "Life is a university—when I have the time at work I study history, politics, religion … I want to change this society, this country … education, health, politics, it's all connected!" he beamed.

Flying Horse, Bloordale: Gentrification for Hipsters by an Immigrant Owner

Flying Horse, a bar on the rapidly gentrifying Bloordale Strip, is an interesting entrepreneurial partnership. The co-manager and owner we interviewed, Thomas, is a young, Canadian-born white man who partners with an Ethiopian businessman, Dabir, who formerly owned the establishment as a single proprietor. Now the co-owner, Dabir, is in charge of cooking the light Ethiopian fare that is served with drinks, while Thomas buys the supplies, serves as DJ, and programs the music.

Thomas described how the neighborhood is changing rapidly. Housing prices all over the city are going up and young families are now buying up houses because they like the location, especially the proximity to the subway line. "Hipsters" are one group that have rapidly moved into the area. He explained how he began his collaboration with Dabir after watching new trendy bars attract busy crowds:

> We are one of the few businesses in the neighborhood that is changing up to new younger crowds where actually the owner is the same owner. You look at [new businesses] that are attracting newer crowds, gentrifiers basically. [Flying Horse] is one of the businesses where actually, the previous owner is benefiting from that [upscaling].

He did not hesitate to use the word gentrification as he talked about these shifts, describing how only a few years ago, "everyone was very Portuguese," but now "hipsters are everywhere."

To attract this younger, "hipster" crowd, the DJ advertises on Facebook, Twitter and a local web site that advertises Bloordale shops and services. He also selects Ontario/Environmentally friendly products to appeal to what he referred to as "different crowds."

Thomas acknowledged that this emphasis on attracting a new crowd has created a divide between new and old clientele. The older folks who used to gather in the bar to drink "cheap beer" and watch a hockey game now come less often. Instead they increasingly hang out in the Chinese restaurant across the street. Thomas struggles with some of these former patrons who can be "very offensive to women:" "We still get older crowds; some of them are dealers, and come inside and keep saying bad things. And we had to ban some of them." At the same time, Thomas emphasized that he gives free food and coffee to a man from the nearby homeless shelter who frequents the bar.

Close Up: Mount Dennis

Changes in the Retail Landscape, 1960–2010

As in Bloordale, changes in the types of businesses in Mount Dennis reflect broader trends in Toronto's retail market structure. As shown in Figure 5, a major shift occurred between 2000 and 2010. Retail services decreased, and personal services grew. A 30 percent decrease in retail services is largely explained by a loss of automobile-related businesses from 10 to 5 establishments. Yet household businesses—home renovation, furniture and upholstery—have remained. On the other hand, with regard to personal services, not only has the number of establishments increased, but the type of businesses has become more concentrated in restaurants and cosmetics, including hair salons and barbershops. These two types make up almost all businesses in the personal services sector.

There are close to 160 local small businesses, approximately 100 of which operate in the Mount Dennis BIA. These small businesses reflect the ethnocultural diversity of the neighborhood. Our survey suggests that nearly 90 percent of the businesses in Mount Dennis are immigrant-owned, while 83 percent are owned by people of color. Two groups predominate: East/Southeast Asian (26 percent) and Afro-Caribbean (23 percent), followed by Africans and Whites.

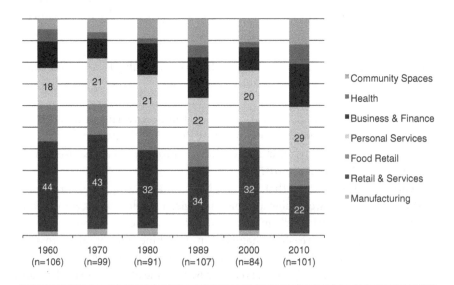

Figure 5   Changes in Retail Businesses, Mount Dennis BIA, 1960–2010

Source: Might Directories (1960, 1970, 1980, 1989); Polk & Co. (2000); Pitney Bowes
   Mapinfo (2010).

The largest proportion of businesses, around 40 percent, has been in operation for less than 5 years. Cross-tabulating with the owners' ethno-racial background reveals that, among recent owners, Afro-Caribbean and East/Southeast Asian immigrants predominate (see Figure 6). Meanwhile, the 30 percent of businesses that have been in operation more than sixteen years are predominantly owned by whites, clearly indicating a process of ethnic succession.

Criminal Insecurity and Social Insecurity

Unlike in Bloordale, gentrification and displacement are hardly concepts that come to mind when working in Mount Dennis. The overall vacancy rate for the Mount Dennis BIA, as determined by a visual survey in December 2011, was 27 percent. This rate is quite high compared to the city average of 9 percent (Cushman & Wakefield 2011). Moreover, the Mount Dennis commercial strip has been characterized by high crime, violence, and intense policing. Mount Dennis was the epicenter of the infamous 2006 "summer of the gun" in Toronto, when six murders were committed in 6 months, allegedly by gangs and youth on Weston Road. According to an interactive map of criminal charges by the police in 2009 and 2010 published in the *Toronto Star*, Mount Dennis and adjacent Weston have crime rates among the highest in the city (Rankin and Winsa 2012).

In response to the high murder rate, Toronto Police initiated its Toronto Anti-Violence Intervention Strategy (TAVIS) in Division 12, which

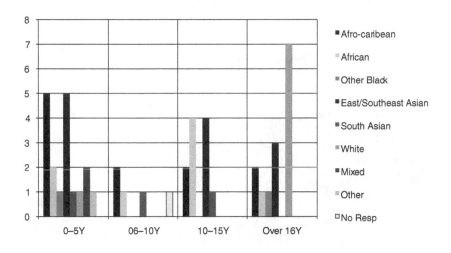

Figure 6 Business Owners' Ethno-Racial Background and Years of Operation in Mount Dennis
Source: Field surveys, 2012.

encompasses the Mount Dennis neighborhood. TAVIS entails temporarily transferring dozens of police officers from less crime-intensive divisions and undertaking a practice of intensive carding in targeted patrol zones. Mount Dennis witnessed a 450 percent increase in the number of cards filled out and filed from 2008 to 2009 (Mutic 2012).

Among shopkeepers, such a heightened level of policing activities and criminal insecurity inspires an intense climate of fear that we experienced firsthand when trying to obtain interviews. As outsiders from the university, we were immediately regarded with suspicion and associated with the law-and-order regime of the local state. We could only gain access through brokering by community-based researchers already known to the shopkeepers or able to identify themselves as area residents. Meanwhile, the climate of fear also shapes characterizations of local businesses as complicit in harboring criminal activity.

In the context of severe social insecurity, reflecting high levels of unemployment, racism in labor markets, concentrated poverty, and urban spatial inequality, it is reasonable to anticipate some points of connection between the inevitable shadow economy and storefront spaces of the commercial street (Venkatesh 2006). Indeed, some bars operate after legal hours, and youth can be seen to congregate in certain bars during the afternoon, "safe" from the surveillance of the street. At the same time, however, outsiders overlook the key roles local businesses play in serving local immigrant populations (see Figure 7). This neglect is, in turn, reflected in planning processes that omit the perspectives, knowledge, and contributions of small business owners.

Figure 7 Local Businesses in Mount Dennis that Serve Immigrant Populations

Source: Photo by Tamam A. Jama.

Innovative combinations of service and retail within a single storefront adapt to the specific needs of immigrant populations and expand our notion of the all-in-one convenience store (Hall 2011). A cell phone dealer provides money transfer services, mailboxes with a fixed address, internet access and fax and copy services, and a hair salon operates a housecleaning service from its storefront. These combinations compare favorably with the usual recommendation that small-scale business should compete with big-box retail by focusing on "specialized products, catering to particular customer needs, high quality service and product knowledge" (Jones and Doucet 2000).

Some businesses in Mount Dennis even attract customers from outside the area. The one most commonly acknowledged by our key informants is an appliance dealer who located in Mount Dennis in 1953. This business has expanded in terms of sales volume, floor area, and market sector, since it now specializes in high-end products, taking advantage of the area's accessibility by major highways. There is also a Korean restaurant that brings in weekly busloads of Korean tourists from New York, Niagara Falls, and Quebec, a second-hand truck dealer who receives business from all over the province, and a vibrant furniture upholstery sector. The latter survives from the days when there were furniture manufacturers in Mount Dennis.

In addition, a solid majority of the businesses we surveyed, 80 percent, indicate that their customers come from both inside and outside the neighborhood. As an immigrant reception area, residential turnover in Mount Dennis is high; we heard frequently that those who are able to establish a regular income move out and "up" to more affluent neighborhoods. We also learned that it is in part through ties to local small businesses that people retain a relationship to Mount Dennis. According to business owners, neighborhood emigrants often return for specific services, like a haircut, from a vendor they know personally and in whose storefront they are likely to see old acquaintances. As on the Lower East Side of New York, long-lasting immigrant-owned businesses and immigrant-oriented services link cycles of immigrant populations who "landed" in Mount Dennis and then moved out, but continue to come back as customers.

The everyday practices of business owners contrast with prevailing representations, particularly those circulating among the proponents of redevelopment planning—of both the "real estate" and "green-cultural economy" variety—who tend to characterize the Mount Dennis commercial strip as an "empty, deficient space." The common perception is that "nothing is going on": "people aren't coming to the neighborhood for anything;" "80 percent of the businesses are vacant;" "our greatest opportunity is that [Weston Road] is a blank slate for anybody to come in and create;" "the Somalis who have moved in, opened variety stores, beauty salons, and go to mosque five times/day, these are the wedges that drive people away;" "there's bars, there's some restaurants, there's some salons, but there's really nothing" (key informant interviews, June 23 to July 19, 2011).

A Redevelopment Narrative: "We Need People with Money"

For proponents of redevelopment, many of them long-time residents of the neighborhood, the shopping street is in decline. The starkest indicator of decline is the concentration of bars, low-end restaurants, barber shops, and hair salons. What is needed, as in the city government's view of Javastraat, in Amsterdam, is to fill the empty storefronts, improve the retail mix, and attract a clientele that will introduce "social mix" into the neighborhood. According to a local green-economy advocate:

> One of the things we need, and it's crass to say, but we need … people who are coming in with money. We just need that balance. Sure, we have our community housing, we have our low-income … You know, we need more of a balance. And that's what we were really hoping … [to] cater to just a little higher end, because whether you're willing to admit it or not, it's the higher end that helps support all the services and stuff that we need for everyone else (interview, key informant, June 23, 2011).

Those engaged in redevelopment planning in Mount Dennis routinely express what our community-based collaborators referred to as "downtown envy"— seeking to remake Mount Dennis in the image of trendy downtown neighborhoods, with visions of flower stands, fresh fruits and vegetables, bakeries, bicycle repair shops, and a characteristic local "brand." As one key informant put it, "residents' associations want to see patios, not cash max and dollar stores; this is about changing retail, the main street, to change the neighborhood" (interview, key informant, July 19, 2011).

However much the "real estate" and "green-cultural economy" redevelopment visions may regard themselves as competing and even ideologically contrary, both are advanced by relatively privileged agents of city building who have the social and cultural capital to leverage opportunities for pursuing their desired future: suburbanization in one sense, gentrification in another. Both entail a fundamental shift in the class composition of the neighborhood, with no explicit regard for displacement pressures that accompany gentrification. Indeed, there is some evidence to suggest that redevelopment initiatives are proceeding in a manner that largely excludes those who currently occupy its commercial spaces.

Today, redevelopment of Mount Dennis is based on a vision of making the neighborhood a destination for outsiders. The vision depicts the shopping street as an empty space, with suboptimal uses, rundown, crime-ridden, and abandoned. But this view pays little attention to how redevelopment will affect existing users.

A Narrative of Social Space: Community Anchors and Mentors

Besides providing essential goods and services to residents, small businesses furnish social space. This function is particularly important in Toronto's inner-suburban neighborhoods that typically lack accessible community spaces, but it is not unlike the similar role played by suburban shopping malls (Parlette and Cowen 2011). Of course, a small business is not a substitute for a community center. Yet it was apparent during our interviews that not everyone was consuming only an economic good or service in the salons, barbershops, computing centers and other shops we visited. People were making social connections.

In a restaurant we might see an old man in a wheel chair, hooked up to a dialysis machine, spending long periods of time, certainly more than it takes to eat a meal, socializing with other seniors, all of them clearly escaping the isolation of their homes. Shopkeepers in Bloordale, both those who predate and those who have "pioneered" the gentrification wave, talk about their role in taking care of the community. They say they want to create spaces that are friendly and responsive to local needs even if customers do not have a lot of money to spend. In Mount Dennis, our community collaborators emphasized how, in the absence of afterschool programs, youth come to local stores and coffee shops to avoid being stopped by police. Reminiscent of Jane Jacobs' (1961) description, local stores serve as a safety net and hub for friendship and relationship building.

Relationships with residents and customers in turn furnish shopkeepers with some security against criminal insecurity. A Vietnamese variety store owner in Mount Dennis explains her approach to crime prevention explicitly in terms of building relationships. To curb once-prevalent shoplifting by youth she does not rely on police or technical security systems. Instead she builds personal relationships with her clientele, including youth. She insists on knowing people by name, and publicly shames them if she catches acts of petty theft. Personal relationships are the best guarantee against shoplifting, she says, and she claims that theft rates in her store are lower than those in the corporate-owned convenience store across the street that relies on security cameras and police surveillance.

Admittedly, some businesses offer places to gather that support anti-social behavior. But there are others with deep commitments to mentoring area youth and steering them away from asocial spaces. Business owners we interviewed acknowledge that they do not have the funds to provide steady employment; most rely on family labor. Few have the capacity to navigate the formal structures of payroll and insurance rules to provide legal jobs. But some nonetheless seek to offer mentorship and job training, like the Mount Dennis barber who provides off-the-books employment to young men who show an interest in the trade, and trains them in barbering skills and business management. In this way he has helped catalyze several new businesses.

A hair salon owner talked with us about her struggles starting her business, which made her want to share her knowledge and experience to help others attempting to enter the same sector. We interviewed a general merchandise trader who aspires to open a gym for youth in the neighborhood; a former Olympics athlete, he wants to provide a structured, supportive environment for area youth to develop an interest in athletics.

In Bloordale, a computer repair service organizes workshops in the local school, a convenience store owner is a prominent leader of an association of Burmese refugees, and the owner of a health food store hopes to coordinate with the four local elementary schools to support snack programs with accompanying nutrition education.

While in most cases, these mentoring practices and relationships do not provide or guarantee formal employment, the literature on immigrant integration and entrepreneurship confirms that they play a crucial role in "overcoming the lack of social and professional networks needed to succeed in the business world" (Wayland 2011: ii, 15). The lack of professional and business mentors "who can 'show the ropes' to a newcomer" is often identified as an obstacle to starting and sustaining a business (Wayland 2011: 15).

Our survey findings indicate that business owners both in Bloordale and Mount Dennis have a higher educational attainment than the wider residential population (and are thus ideally suited to a mentoring role). Cross tabulating educational background with immigrant status and age of business reveals that educational qualifications are concentrated among immigrant business owners, particularly recent immigrant business owners operating their business less than 5 years. This trend corresponds to the findings of a recent report on immigrant entrepreneurship by two Toronto-based private foundations, who find that recent immigrants generally have high levels of education and experience high levels of involuntary self-employment (Wayland 2011: 15). We encountered new immigrant small business owners with degrees in political science, engineering, accounting, and computer IT, who had resorted to business after facing racism and non-recognition of their credentials in the labor market.

All of these practices and relations exemplify the positive social functions performed by small businesses in low-income neighborhoods. These rarely surface in stories of neighborhood decline and transformation, but they are an important element of the cultural ecosystem of local shopping streets.

Business Improvement Area or Community-Based Initiative?

At the neighborhood scale, the official, municipally mandated form of business organization is the BIA. In Bloordale, our field survey suggests that almost 80 percent of businesses are aware of the local BIA. However, the awareness does not necessarily translate into participation in BIA activities. Business owners commonly express considerable caution and suspicion toward the BIA, in view of the Board of Management being chaired by a strip club owner and the lead role being played by a resident artist who does not own a storefront business and is perceived (however inaccurately) to be promoting her own agenda.

The potential for arts-led improvements to promote perceptions of safety and indeed property value increases is not lost on the Board Chair, but many businesses question how a focus on aesthetics could benefit their business. The arts-focused annual street festival is widely regarded as featuring artists and small entrepreneurs from outside the neighborhood, while actually disrupting regular weekend shopping routines. Some of the newer businesses have started to coordinate informal networks of support for sourcing goods and services, building up a community of business exchanges, and have considered taking these initiatives into a more formal involvement with the BIA ("taking it over," as some put it). Thus far, the initiative has involved newer, gentrifying business owners in an informal capacity, and has not extended to the older low-income and immigrant business owners.

In Mount Dennis, the BIA is essentially unknown to most business owners; 70 percent in our sample were not aware of the very institution charged with representing and supporting local businesses, much less the municipal programs accessible through BIAs that are intended to support small neighborhood-based businesses. In addition, some business owners we interviewed feel socially isolated and incapable of navigating bureaucratic processes to access municipal programs and services.

Like the Bloordale BIA, the Mount Dennis BIA is also managed by a small core of influential individuals who are not deeply embedded in the social spaces of the commercial street. In this case the BIA is led by a powerful local city councilor who routinely enrolls the BIA in the "real estate" vision of redevelopment. Levels of participation of business owners in BIA activities are extremely low. However, an alternative business network has recently developed in Mount Dennis, as well. The initiative emerged out of our research process and has achieved a greater degree of institutionalization than the initiative we learned about in Bloordale.

The West End Local Economic Development (WE-LED) group was formed by our community-based collaborators who have been involved with interviewing business owners and seek to develop linkages between local businesses and youth in Mount Dennis (much as advocated by the service agencies we interviewed in Bloordale). The expectation is that this kind of organizing will both advance mentoring and employment opportunities for youth in the neighborhood while also informing especially immigrant-owned businesses about redevelopment planning processes so that they might interject their own perspectives and visions for the neighborhood. It is important to emphasize that the formation of WELED has been facilitated by the Action for Neighbourhood Change office, a core element of the social infrastructure formed through the Priority Neighbourhood targeted poverty reduction strategy.

In contrast, such social infrastructures are absent in Bloordale, where alternative modes of business organizing are led instead by recently arrived high-end businesses. In both cases, planning for commercial spaces continues to be orchestrated by those with political and social forms of capital necessary to access the municipal bureaucracy, private sector investors as well as local constituencies for neighborhood improvement. The perspectives of business owners providing affordable goods and services on disinvested local shopping streets, particularly those who are newly immigrated, continue to be excluded from planning for the future.

Shopkeepers' Stories

Bauer's Treasure, Weston/Mount Dennis: An Unstable Neighborhood

This specialty store is located on the Weston Road thoroughfare that forms the commercial strip in Mount Dennis and neighboring Weston Village. Located just North of the Mount Dennis strip itself, Bauer's Treasure is featured because of the active role the store owner has played in planning meetings and workshops addressing redevelopment of the nearby Kodak lands and the "mobility hub" associated with planned transit infrastructure. Packed with an eclectic, colorful assortment of specialty goods carefully displayed old wooden tables and shelves the store has a cheerful atmosphere that detracts attention from its state of physical decline—cracked ceilings, signs of mold and water damage.

The shopkeeper's father, who the owner, Stella, described as an "upper middle class, white Jewish man," bought the small store back in 1980 when Weston Village was what she described as a "very stable neighborhood." She attributed this stability to the fact that central core of the neighborhood was home to what she called a "homogeneous and stable" community made up families who had lived in the area for four or five generations. Their history in the neighborhood, she claimed, created a strong sense of community and a "multigenerational" aspect to running her business

Stella also claimed that the thousands of manufacturing jobs that existed in the neighborhood including the CCM Bicycle Factory, Moffett Appliances and the massive Kodak plant in Mount Dennis contributed to her business's early success. However, small businesses in the area have struggled since manufacturing jobs started to decline. One by one, the big factories shut down. And as the factories closed due to external pressures, people stopped doing their errands along the area's commercial thoroughfares. She described this as a "trickle-down effect" that negatively impacted Weston Road.

She also described how, around the same time, a shopping mall featuring a liquor store, the Canadian Tire chain (hardware and automotive supplies), a bank and a chain grocery store opened nearby. For her, the loss of manufacturing jobs and the new mall meant the demise of small commercial spaces in her neighborhood. She said, "Workers used to bring their wives and children to shop for special occasions; people went to restaurants, fab dress and shoe shops that had been here for fifty years … we relied on foot traffic … then the mall was like a nail in the coffin for retail."

Stella described with unease the cultural transitions that have transpired on the strip as immigrant populations, particularly from East Africa, moved into the neighborhood. For her, these new populations do not represent an increase in foot traffic. Rather she perceives them as weakening the neighborhood's formerly "stable" identity.

Tranquility Bar, Mount Dennis: A Working Man's Bar

Jack is the owner-manager of a long-time bar that fit the description of what key informants involved in redevelopment planning would call a "hole in the wall." In contrast to the bright June early afternoon sun, inside it was dark; old, fluorescent lights flickered, the Formica tables were chipped and the air felt stale and heavy.

An older Trinidadian man, Jack appeared uncomfortable being interviewed by a university-based, white research assistant carrying a laptop. When first asked how he had started his business, he said tersely, "I can't share personal details." The accompanying community-based researcher, also Trinidadian, then jumped in to help break the ice, pointing out that when he first moved to Canada he had established many community connections and friendships playing dominos with the men who hang out in this bar.

Once he heard this appreciation and recognition, Jack went on to acknowledge how the bar plays an important social role in Mount Dennis. When we asked him who his customers were, he said in a straightforward tone "in my business, nobody special, old men, working men, nobody special." He then went on to explain that most of his clientele are "lonely and bored, older men" who like to sit in the bar for hours and play cards. Other clientele are working men who "play dominos after a hard day of work." A quick glance around the bar told a story about the social role the bar played; one older man sat with his portable dialysis machine, drank a can of coke and played cards with friends.

As we probed a bit more with questions, Jack explained how he runs some of his business on trust and networks. He described how sometimes customers arrive without enough money to pay for their drinks, but if he knows them and trusts them he lets them run a tab. After sitting in Jack's bar for about an hour we were impressed by the significance of the small bars and coffee shops in Mount Dennis for poor and newcomer seniors in particular. Because the neighborhood lacks supports for seniors—a pharmacy, decent community center space, reliable public transportation—such spaces furnish de facto community gathering places, even as they are commonly disparaged by planners, politicians and BIA leaders as the empty and deficient spaces to be revitalized through redevelopment.

The Future of Toronto's Shopping Streets

Evidence of gentrification in Toronto to date—whether through the arrival of high-end businesses in Bloordale or exclusionary planning processes in Mount Dennis—suggests that affordable commercial spaces will not be protected without organized community response or planning intervention. Downtown, where average household income has increased at a faster rate than in the rest of the city for fifty years, gentrification has rendered the core of the city highly inaccessible to low-income communities; as one of the last holdouts of affordability, Bloordale has predictably been staked out by the art galleries and start-up boutiques.

Debates over the merits and costs of gentrification circulate in the public realm, and many new shopkeepers express a desire to retain affordability and social mix and avoid the homogenization that accompanies gentrification in surrounding neighborhoods. At the same time, they have embraced law-and-order regimes that manage criminal insecurity with punitive measures, and fail to engage the local service agencies providing supports to marginalized and precarious local populations.

The relationship to police surveillance on an inner-suburban, racialized, low-income shopping street is more ambiguous. Shopkeepers, many of them new immigrants, clearly require protection from crime just as anyone does, but the racialized nature of police carding is widely regarded as discriminatory. Shopkeepers may themselves have experienced the arbitrary power of the state through mundane local planning measures such as liquor licensing and parking enforcement that are widely perceived to disadvantage new immigrant business owners. Processes of redevelopment planning have excluded shopkeepers, overlooked the vital services they perform for surrounding immigrant communities in particular, and advanced revitalization scenarios that will likely generate the same kind of displacement pressures currently experienced in gentrifying downtown neighborhoods.

Narratives of everyday life on both streets expose common experiences of precarity: with immigrants and low-income users of the space, whether business owners or customers, experiencing lack of official recognition for their professional credentials, exclusion from formal planning processes combined with subjection to formal planning regimes, and an ambiguous relationship with criminal insecurity. At the same time, we have seen that shopkeepers play key roles creating social space and making the city accessible to low income and new immigrant populations. Yet these roles are not recognized by urban planners, and they have grown increasingly fragile, not only in the rapidly gentrifying downtown core, but even in the inner suburbs experiencing redevelopment pressure.

What is at stake, then, in the future of Toronto's shopping streets, is the role of commercial space in securing the right to the city. Interjecting a politics of class and race helps to counter common-sense narratives that depict commercial space as the terrain of an abstract, disembodied, competitive market—and foregrounds instead the institutions and human decision-making behind the way markets are organized. Doing so opens up possibilities to decide what kind of commercial spaces might contribute to forging the just city. Local shopping streets in Tokyo face this question in a different way.

Acknowledgements

This chapter is based on semi-structured interviews with local business owners in Mount Dennis and Bloordale, each conducted by a team of researchers drawn from Rankin, Kamizaki, McLean, and other students. The Mount Dennis research team included community-based researchers, who paired with the university-based researchers for each interview. The researchers also carried out a quantitative analysis of the residents in each area and the changing types of businesses on each street, using Statistics Canada census data from 1971 to 2006, property assessment rolls available at City Hall, and business entries in Might's Street Directory and Canada Business Data. Finally, we conducted a short survey on demographic and business characteristics in both shopping streets, supplemented by media research and policy document reviews, as well as two rounds of key informant interviews with city planners, real estate agents, and community leaders. Thanks go foremost to Cutty Duncan for collaborating with us in this research and leading the team of community-based researchers in Mount Dennis. They also go to Kurt Strachan, Sojica John, Gail Matthews, Shane McLeod, Ayana Francis, Michael Miller and Fatma Yasin who contributed as community-based researchers. Shopkeepers' names have been changed to protect their identity.

Note

1   These street designations refer to the names of commercial strips and BIAs traversed by Weston Road and Bloor Street respectively; the streets themselves are very long and encompass numerous commercial strips, so for demarcation purposes, the popular strip and BIA names have been used here.

References

Artscape. 2011. "Cultural/Creative Hubs in Priority Neighbourhoods: Feasibility Study for a Cultural/Creative Hub in Weston Mount Dennis." Available at www.toronto.ca/legdocs/mmis/2012/ey/bgrd/backgroundfile-45013.pdf (accessed April 20, 2012).

Brown-Saracino, Japonica. *A Neighborhood that Never Changes: Gentrification, Social Preservation and the Search for Authenticity.* Chicago IL: University of Chicago Press.

City of Toronto. 2012. "Planning Applications." Available at http://app.toronto.ca/DevelopmentApplications/searchPlanningAppSetup.do?action=init.

City of Toronto. 2013. "Section A: BIA overview—Toronto BIA Operating Handbook." Available at www1.toronto.ca/wps/portal/contentonly?vgnextoid=bc13ad51a40ea310VgnVCM10000071d6 0f89RCRD&vgnextchannel=ea3032d0b6d1e310VgnVCM10000071d60f89RCR (accessed January 16, 2014).

Cushman & Wakefield. 2011. *MarketBeat Retail Snapshot: Toronto/Canada*. A Cushman & Wakefield research publication, quarter 4. Available at www.cushwake.com/cwmbs4q11/PDF/retail_toronto_4q11.pdf (accessed March 15, 2012).

Galabuzi, Grace-Edward F. 2007. "Marxism and Anti-racism: Extending the Dialogue on Race and Class." *Marxism: A Socialist Annual* 5: 47–49.

Goonewardena, Kanishka, and Stefan Kipfer. 2005. "Spaces of Difference: Reflections from Toronto on Multiculturalism, Bourgeois Urbanism and the Possibility of Radical Urban Politics." *International Journal of Urban and Regional Research* 27(3): 670–78.

Gupta, Rahul. 2009. "Bloor and Lansdowne's Hidden Side." *Creative City: Art Transforming Toronto*. December 8. Available at http://thecreativecity.wordpress.com/2009/12/08/bloor-and-lansdowne%E2%80%99s-hidden-side (accessed December 11, 2014).

Hackworth, Jason, and Josephine Rekers. 2005. "Ethnic Packaging and Gentrification: The Case of Four Neighborhoods." *Urban Affairs Review* 41(2): 211–36.

Hall, Suzanne M. 2011. "High Street Adaptations: Ethnicity, Independent Retail Practices, and Localism in London's Urban Margins." *Environment and Planning A* 43: 2571–88.

Hall, Suzanne M. 2012. *City, Street and Citizen: The Measure of the Ordinary*. London: Routledge.

Harris, Richard. 1999. *Unplanned Suburbs: Toronto's American Tragedy, 1900 to 1950*. Baltimore MD: Hopkins Fulfillment Service.

Hershberg, Erin. 2011. "Gentrification with a side of co-operation." *The Globe and Mail* (February 26): M7.

Hulchanski, David. 2010. *The Three Cities within Toronto: Income Polarization among Toronto's Neighbourhoods, 1970–2005*. Toronto: Cities Centre, University of Toronto. Available at http://3cities.neighbourhoodchange.ca (accessed March 8, 2015).

Ireland, Carolyn. 2010. "Bloordale: A Vintage Draw for First-Time Buyers." *The Globe and Mail* (October 28): G16.

Jacobs, Jane. 1961. *The Death and Life of Great American Cities*. New York: Random House.

Jones, Ken, and Michael Doucet. 2000. "Big-box Retailing and the Urban Retail Structure: The Case of the Toronto Area." *Journal of Retailing and Consumer Services* 7: 233–47.

Kaminer, Michael. 2012. "An Art Scene Blooms in a Toronto Neighborhood." *New York Times* (September 9). Available at www.nytimes.com/slideshow/2012/09/09/travel/20120909-SURFACING.html (accessed January 16, 2014).

Lo, Lucio. 2009. "Immigrants and Social Services in the Suburbs." In *In-Between Infrastructure: Urban Connectivity in an Age of Vulnerability*, edited by D. Young, P. B. Wood, and R. Keil, pp. 101–14. Vancouver: Praxis E-press.

Marshall, Sean. 2013. "A Brief History of Dufferin Street." August 5. Available at http://spacing.ca/toronto/2013/08/05/history-dufferin-street (accessed January 16, 2014).

Mazer, Katie, and Katharine Rankin. N. 2011. "The Social Space of Gentrification: Limits to Neighborhood Accessibility in Toronto's Downtown West." *Environment and Planning D: Society and Space* 29(5): 822–39.

McBride, Jason. 2008. "Redeeming Bloordale, One Gallery at a Time." *The Globe and Mail* (June 21): M.5.

McGinnis, Rick. 2001. "Happy Days: An Odd Jumble of Small Homes North of the Junction, Mount Dennis Doesn't Have the Kind of History You Find in Guidebooks." *Toronto Life* (July).

McKittrick, Katherine, and Clyde Wood. 2007. *Black Geographies and the Politics of Place*. Brooklyn NY: South End Press.

McLean, Heather. 2009. "The Politics of Creative Performance in Public Space: Towards Toronto Case Studies." In *Spaces of Vernacular Creativity: Rethinking the Cultural Economy*, edited by T. Edensor, D. Leslie, S. D. Millington and N. Rantisi, pp. 200–214. London: Routledge.

Might Directories. 1960. *Might's 1960 Greater Toronto City Directory*. Toronto: Might Directories.

Might Directories. 1970. *Might's 1970 Metropolitan Toronto City Directory*. Toronto: Might Directories.

Might Directories. 1980. *Might's 1980 Metropolitan Toronto City Directory*. Toronto: Might Directories.

Might Directories. 1989. *Might's 1989 Metropolitan Toronto City Directory: West Edition*. Toronto: Might Directories.

Monsebraaten, Laurie. 2009. "Last chance for Weston, Toronto's rustbelt." *Toronto Star* (April 26): A10.

Mount Dennis Community Association. 2007. "Lost in Translation." Presentation given at the People Plan Toronto Summit, Toronto, ON.

Murray, Whyte. 2008. "Where Art Goes, Others Follow: West Queen West's Art Moves North". *Toronto Star* (October 5): E1.

Mutic, Anja. 2012. "One Day in Toronto's Coolest Neighbourhood." Available at www.lonelyplanet.com/canada/travel-tips-and-articles/76078 (accessed January 16, 2014).

Parlette, Vanessa, and Deborah Cowen. 2011. "Dead Malls: Suburban Activism, Local Spaces, Global Logistics." *International Journal of Urban and Regional Research* 35(4): 794–811.

Pitney Bowes Mapinfo. 2010. Canada Business Data. New York: Pitney Bowes Mapinfo.

Polk & Co. 2000. *Toronto/Central West York Polk Criss-Cross Directory.* Livonia MI: R. L. Polk & Co.

Rankin, Jim, and Patty Winsa. 2012."Known to Police: Violent Crime in Weston-Mount Dennis is Down, Youth Feel Harassed by Toronto Police." *Toronto Star* (March 9).

Rankin, Katharine N. 2008. *Commercial Change in Toronto's West-End Neighborhoods.* Research Paper 214. Toronto: Cities Centre, University of Toronto.

Rankin, Katharine N., and Jim Delaney. 2011. "Community BIAs as Practices of Assemblage: Contingent Politics in the Neoliberal City." *Environment and Planning A* 43: 1363–80.

Rankin, Katharine N. and Heather McLean. Forthcoming. "Governing the Commercial Streets of the City: New Terrains of Disinvestment and Gentrification in Toronto's Inner Suburbs." *Antipode.*

Rankin, Katharine N., Kuni Kamizaki, and Heather McLean. 2013. *The State of Business on Weston Road: Disinvestment and Gentrification in Toronto's Inner Suburbs.* Research Paper 226. Toronto: Cities Centre, University of Toronto. Available from: www.citiescentre.utoronto.ca/publications.htm (accessed March 8, 2015).

Roberts, David and Minelle Mahtani. 2010. "Neoliberalizing Race, Racing Neoliberalism: Representations of Immigration in the Globe and Mail." *Antipode* 42(2): 248–57.

Simmons, Jim. 2012. *The Evolution of Commercial Structure in the North American City: A Toronto Case Study.* Toronto: Cities Centre, University of Toronto.

Toronto Life. Undated. "Real Estate Guide: Mount Dennis." Available at www.torontolife.com/real-estate-guide/west/mount-dennis (accessed January 18, 2014).

Venkatesh, Sudhir Alladi. 2006. *Off the Books: The Underground Economy of the Urban Poor.* Cambridge MA: Harvard University Press.

Viswanathan, Leela. 2010. "Contesting Racialization in a Neoliberal City: Cross-cultural Collective Identity as a Strategy among Alternative Planning Organizations in Toronto." *GeoJournal* 75(3): 261–72.

Wacquant, Loïc. 2007. *Urban Outcasts: A Comparative Sociology of Advanced Marginality.* Cambridge: Polity.

Wayland, Sarah V. 2011. *Immigrant Self-employment and Entrepreneurship in the GTA: Literature, Data and Program Review.* Toronto: Metcalf Foundation and Maytree Foundation.

Zukin, Sharon. 2009. "Changing Landscapes of Power: Opulence and the Urge for Authenticity." *International Journal of Urban and Regional Research* 33(2): 543–53.

Zukin, Sharon. 2010. *Naked City: The Death and Life of Authentic Urban Places.* New York: Oxford University Press.

Zukin, Sharon, Valerie Trujillo, Peter Frase, Danielle Jackson, Danielle Recuber, and Abraham Walker. 2009. "New Retail Capital and Neighborhood Change: Boutiques and Gentrification in New York City." *City and Community* 8(1): 47–64.

CHAPTER 7

Tokyo's "Living" Shopping Streets

The Paradox of Globalized Authenticity

KEIRO HATTORI, SUNMEE KIM,
AND TAKASHI MACHIMURA

Local shopping streets in Japan are threatened by the same factors that put small retail businesses in danger all over the world. Economic uncertainty limits consumers' ability to spend, and individual owners face serious competition from transnational and domestic chain stores, mega-supermarkets, and online shopping. Throughout Japan, shopping streets in small cities are reeling from these pressures. Even in Tokyo, the capital, a city with more than 12 million residents, the number of small retail stores with fewer than five workers fell from 93,000 in 1997 to 63,000 in 2007 (Tokyo Metropolitan Government 2001, 2009).

These changes shape the dominant Japanese view that local shopping streets are in decline. This view is reinforced by the conservative attitude of many shopkeepers, and especially by the conservative leadership of shopkeepers' associations on individual streets, some of whom manage family businesses that have lasted for two or three generations.

But recent events suggest a dramatically different view. On a Sunday afternoon in April, 2011, not long after the disastrous tsunami, earthquake, and explosion at the Fukushima nuclear power plant, a crowd of thousands thronged the narrow shopping street of Koenji, a Tokyo neighborhood that has been known since the 1960s as a mecca for underground, avant-garde youth culture, to protest Japan's dependence on nuclear power.

The demonstration was not announced in advance. Even the police, who are always diligent in dealing with demonstrations, were caught off guard and were unable to cope with its scale. The event also received little coverage in the mass media because most of the large newspaper and broadcasting companies were not prepared to cover it. Yet it was the first large-scale street protest in Tokyo following the earthquake.

The unlikely organizer of this protest demonstration was Matsumoto Hajime, the manager of a second-hand lifestyle and home décor shop in Koenji (Tosa 2011). He was joined by other young people who had opened second-hand clothing shops, as well as furniture stores, bars, and cafés, in the surrounding area in the early 2000s. All the organizers promoted the demonstration through social media. Their shops, and the public space of the streets around them, offered a ready-made "scene" for protest.

The owners of this loose cluster of shops in Koenji, who formed a group known as "the Amateurs' Revolt," make the point that, in recent years, local shopping streets in Japan have gradually developed the image of a space that symbolizes "vitality" and "vigor." Television networks and magazines often run special features on shopping streets that emphasize their enduring charm. Moreover, in Japanese city-planning policy, shopping streets are often considered to be a historical and cultural landscape that represents a typical Japanese character.

Events like the protest against nuclear power suggest that shopping streets have become screens on which Japanese people project the imagination of a new era of national development. Though they may look traditional, local shopping streets are really shaped by people's hopes and dreams for the future. Yet this aspirational image of the space contrasts with its sometimes dismal economic performance.

What is the role, then, of Tokyo's local shopping streets, in the face of the mega-malls of global capitalism? Are these streets a form of local resistance to globalization? Or are they simply a familiar space of traditional Japanese cultural values, where the kimono shop, the baker of *sembei* (Japanese crackers), and the *soba* (buckwheat noodle) restaurant offer a convenient place for escapist nostalgia?

Perhaps all three views are partially correct, for local shopping streets in Tokyo are rich in traditional aesthetics but flexible in the products and services they offer. They are historic and trendy, Asian and Western, and in that paradoxical mix of globalized authenticity they find their ability to survive.

Two Shopping Streets: Azabu-Juban and Shimokitazawa

In this chapter on Tokyo, we will visit two shopping districts with contrasting geographical locations and historical backgrounds (see Figure 1). One is Azabu-Juban, a main shopping street with a few commercial side streets which

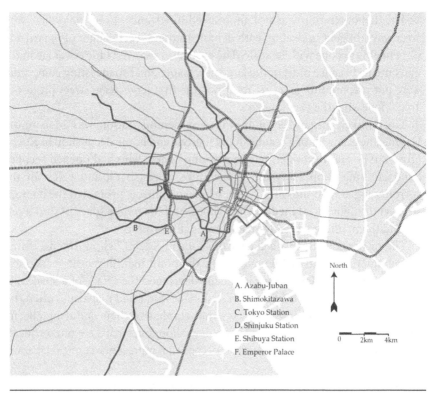

North

A. Azabu-Juban
B. Shimokitazawa
C. Tokyo Station
D. Shinjuku Station
E. Shibuya Station
F. Emperor Palace

0   2km   4km

Figure 1   Tokyo, Showing Locations of Azabu-Juban and Shimokitazawa
Source: Map drawn by Keiro Hattori.

is located just outside Tokyo's city center (i.e. areas C and F in Figure 1) and is accessible, since the turn of the twenty-first century, by the Namboku and Oedo subway lines. Azabu Juban has had a long history as a shopping street since the feudal Edo period (1603–1868). Even now, many small retail businesses, some of which opened their doors almost two hundred years ago, sell foods and daily goods to local residents. But on the same streets, and particularly in the adjoining alleys, shoppers looking for high-quality goods find them in tiny, sophisticated, and often costly restaurants and shops.

The contrasting shopping district is Shimokitazawa. It is located in a western suburb of Tokyo at the intersection of the Odakyu and Keio Inokashira railroad lines, a short ride from the city center. Built on farmland before the era of the automobile, the complex, irregular tangle of narrow streets and alleys makes it quite difficult for cars to drive into the shopping area. Some businesses sell goods and services that meet the everyday needs of local residents, from grocery and hardware stores to barber shops. But there are also many boutiques, bars, hair salons, and restaurants that cater to hipsters.

The types of shops in these two districts are quite distinctive, but also different. Azabu-Juban is known for upscale fashions and foods, while Shimokitazawa features hip music bars, small theaters, and vintage clothing boutiques. In contrast to the elegant French pastry shops that sell *macarons* in Azabu-Juban, a takeout shop in Shimokitazawa sells doughnuts made with *natto*, a fermented soybean paste that many Japanese admit is an acquired taste. At the same time, these streets share several characteristics which reveal how vibrant the social life of a Tokyo shopping street can be.

First, both Azabu-Juban and Shimokitazawa are exceptional survivors. In contrast to some local shopping districts in the city, they have been able to compete with mega-shopping malls and global chains. The secret of their success cannot be the way they look. The streets of both Azabu-Juban and Shimokitazawa look very much like other small Japanese streets.

However, if you look more carefully, you will recognize common charms. Both districts maintain a very specific ambiance or milieu, deeply rooted in their local contexts. Both are very "Japanese," but this does not mean that they are limited to a traditional East Asian décor and atmosphere. Instead, the dominant aesthetic of both streets is "Japanese" in the sense that Japan has always freely mixed, borrowed, and blended elements of foreign cultures.

In Azabu-Juban, Western influence takes the form of French restaurants, artisanal chocolate shops, and even a stylish café, Hudson Market Bakers, inspired by the owner's experience while living in New York. Yet becoming "Western" is not the only option. Other shops imitate "Korean" or a selective "Asian" character, and even recreate "forgotten Japanese." All of these formats compete and co-exist in the same local shopping street.

Many businesses in Shimokitazawa are equally influenced by the West, but instead of upscale sophistication, the vintage clothing stores are like those found in New York's East Village, and the music bars are like those in Brooklyn's Williamsburg. While a new wine bar may open in Azabu-Juban, a remarkable new business to open in Shimokitazawa and to be discovered by food bloggers is the *natto* doughnut shop.

Shoppers come to Azabu-Juban and Shimokitazawa to enjoy the contradictory sense of authenticity that these streets imply. They want to consume not just tangible things, but also an atmosphere of "nostalgia," "resilience," or even "resistance" to a widespread sense of economic uncertainty and cultural loss. They want to consume forms of intangible heritage such as local history and memory. Both shopping streets survive because they manage to keep creating, more or less, the feeling of local authenticity that shoppers desire. Authenticity has become their commercial "niche."

However, in both streets, the special qualities of local authenticity confront the homogenizing forces of global capitalism. When local authenticity is discovered by the mass market, it becomes a common commodity. "Living" "authentic" local shopping streets like Azabu-Juban and Shimokitazawa risk becoming unpopular if authenticity is routinely repeated in many stores selling the same kinds of things, or, ironically, if the pace of transnational, multicultural innovation slows down.

Azabu-Juban

Streetscape

Racks of women's clothes, towels, and lingerie line the front of the Western-style Nishimoto shop, in the center of Azabu-Juban's main street. Chain stores on both sides are painted in pale, quiet colors that match the calm Juban atmosphere. A woman dressed in a jogging suit and sunglasses walks her dog. She pauses to rest on the terraced seating outside Starbucks. Above each shop are several floors of either offices or apartments.

Figure 2   Azabu-Juban, Shop Selling Karinto, a Traditional Fried Sweet Pastry from Okinawa
Source: Photo by Sharon Zukin.

At first, relatively high-rise, glass-walled condominiums attract your attention. Yet, walking down the street, you find a lot of old, low-rise buildings. A plaque on one building's façade marks the 100th anniversary of the small shop on the ground floor. Longtime shop owners who have decided to stay in Juban have rebuilt their old wooden shops as new, modern buildings, particularly after the removal of the traditional pedestrian arcade in 1991. But even now, the dominant look of the street is a cozy mixture of building styles and sizes.

You can gaze out at the shopping street from the second-floor window of McDonald's. A Caucasian man on a bicycle and some businessmen rush by. From the opposite end of the street, a Caucasian mother pushes her young daughters in a stroller; the girls are already dressed in costumes for Halloween, a few days away. Toward the end of the road, you see large signs for the pawn shops, hair salons, restaurants, and bars which line the street.

With a gust of wind, the smell of *yakitori* (grilled pork) wafts down from the direction of Abe-chan, the neighborhood's well-known *izakaya* (a bar–restaurant or pub) that for the past forty years has been famous for its special sauce. You are aware of a delicate fusion between the traditional smells and sights of a Japanese shopping street and its Western feel.

A Historical Palimpsest: One Layer of Time on Top of Another

In the mid-1980s, billboards were erected on the roads near Azabu-Juban that announced a strange slogan: "Tibet resolution in Minato." In Japanese, "Tibet" is a figure of speech that refers to a region that is difficult to get to by public transportation. The central core of Tokyo is divided into 23 wards, and Minato, in which Azabu-Juban is situated, is at the very geographical center. Yet Azabu-Juban, which occupies the lowland along a river and is surrounded on three sides by hills, has historically been isolated from the rest of the city. In this sense, Azabu-Juban for a long time was Tokyo's "Tibet."

Before the first subway station was built at one end of the main shopping street in the early 2000s, it took almost thirty minutes to walk to the nearest subway line. But because of its extremely inconvenient location, Azabu-Juban was able to retain architectural and commercial elements from earlier eras. They easily coexisted on the shopping street, and gave it a specific, traditional Japanese character.

Azabu-Juban first took form as a street in the Edo period, when it was built in front of the gate of Zenpukuji, a Buddhist temple founded in AD 824. Many people continued to gather at Azabu-Juban during the Meiji period (1868–1912), when it was a shopping district catering both to pilgrims to the temple and to wealthy residents who lived on the surrounding hillsides. Some shops that are still on this street today have been in business there since that time.

Due to increasing numbers of cars, city trams were abolished during the 1960s and 1970s, leaving Azabu-Juban with no access to public transportation.

Although this was a disadvantage for attracting customers from outside the immediate area, Azabu-Juban came to occupy something of a geographical niche.

The district also enjoys a cultural niche. Despite its outwardly Japanese appearance, Azabu-Juban has long been influenced by direct interactions with the West. These can be traced back as far as 1859, when the U.S., after pressing Japan to open its markets to Western nations, established its first diplomatic facility in Tokyo in Zenpukuji. Following the Meiji period, a large number of foreign embassies were built on the surrounding hills. For this reason, many foreigners became neighborhood residents. To serve them, shops selling imported goods and supermarkets were quickly established in Azabu-Juban.

During the boom years of the 1980s, when Japan was rejoicing in its global economic success, Azabu-Juban was "discovered" as an *anaba* (a little-known, good place) for shops and clubs. Located near Roppongi, a district known for evening amusements, Azabu-Juban in 1984 became the home of Maharaja, a high-class disco that symbolized the frenzy of the speculative, "bubble" economy. This club helped to develop Azabu-Juban's image as a high-class commercial destination. But, in the second half of the 1980s, the street faced a crisis, due to a rapid rise in housing prices and the subsequent collapse of the bubble economy.

Then, in the early 2000s, when the first subway station was built, the shopping street was struck by a new wave of capital investment. Three years later, Roppongi Hills, an enormous, high-rise, mixed-use development was built nearby, with luxury shops and condominium apartments, high-class restaurants, offices for financial firms, movie theaters, and an art museum. This strengthened the upscale image of the entire district.

The interactions between history and geography have created in Azabu-Juban a special balance of "globalized authenticity." The streetscape is imbued with a nostalgic "Japanese-ness," the products and people evoke the West, and the high-class image is a memory of the bubble era: all these layers create Azabu-Juban's "living" character.

But Shimokitazawa developed in a very different way.

Shimokitazawa

Streetscape

Getting off the train at the Shimokitazawa station, you see an unusual, big, empty space in front of a bustling commercial district. The space has been cleared for redevelopment, but Kawai, a bar, still remains. Its owner has insisted on keeping the bar open until the last possible day.

Inside Kawai, you pay only 2000 yen ($20) and you can drink as much as you want, but the choice of alcoholic beverages is not very good. You cannot order special dishes because the bartender prepares his own small *otoshi* (appetizers) regardless of what the clientele may prefer. Customers come and go, chatting with the bartender and with each other. Many say the redevelopment of the area in front of the railroad station will ruin Shimokitazawa's special character.

Walking in Shimokitazawa is quite safe even if you are drunk. This is because you rarely encounter cars. In fact, there is not a single traffic light around the station. The streets are very narrow, and many of them are dead-end. This is a good example of the human scale of the urban environment in most traditional Japanese cities, creating what Chester Liebs, an American scholar of cultural landscapes, calls a "bicycle neighborhood" (Liebs 2011; see Figure 3).

To the north of the station, there are many clothing shops, cafés with foreign tastes, and sophisticated but reasonable French or Italian restaurants that cater mainly to young women. To the south, however, there are many bars, music clubs, and second hand clothing shops that cater mainly to young men. Here there are also transnational and national chains, from McDonald's and Uniqlo to Gyoza no Ohsho and Mister Donut. This is a somewhat wilder retail district than the more sophisticated northern zone.

Many bars stay open until the wee hours of the morning. Legendary musicians like Akira Sakata, Kazutoki Umezu, and Natsuki Kido perform at the jazz

Figure 3   Small Shops in Shimokitazawa
Source: Photo by Keiro Hattori.

bar Lady Jane, where a whiskey bottle once bought by one of the most charis-matic Japanese movie stars, the late Yūsaku Matsuda, is kept on display. Yutaka Oki, the bar owner, says that an "agglomeration economy" helps the music bars and clubs that cluster together in the neighborhood.

Several of the bars are famous. Chizuko Yamazaki, the owner of the rock bar Mother for more than forty years, was depicted in a best-selling novel by Banana Yoshimoto (2010). A regular customer who is now 45 years old says she has been coming here since she was 19, when she read about the bar in maga-zines.

Yamazaki is friends with Carmen Maki, a legendary rock singer who was a charismatic figure in the 1970s. Maki is also a Mother customer. "In Shimokitazawa," Yamazaki says, "there is no gap between regular people and celebrities. Here, people are the same." If you heard this phrase outside of Shimokitazawa, you would laugh. But here, there is an air of hippie-like opti-mism reminiscent of the 1960s and 1970s that makes you think it may be true.

History: From Farming Village to Epicenter of Youth Culture

Despite its somewhat chaotic streetscape, Shimokitazawa is one of Tokyo's most vibrant shopping districts. It is known as a capital of "hipster" and "under-ground" culture.

The area was developed as a residential district after the Great Kanto Earthquake of 1921. During World War II, the area miraculously escaped Allied bombing, which enabled it to keep its traditional buildings and cultural char-acter. After the war, many tiny shops clustered near the railway station selling scarce goods that came through the informal channels of the black market (this is the area that was cleared for redevelopment in 2013). In the following years, so many retail shops opened that they began to encroach on the residential area, a disorderly pattern that continued without much intervention by the city government.

As a consequence, streets remain narrow, and there are only a few high-rise buildings. These qualities are rare in Tokyo's business centers, but, like the lack of public transportation until recently in Azabu-Juban, they create a unique positive attraction in Shimokitazawa.

During the 1960s, the area south of the railroad station began to see some red-light activities catering to businessmen. But after the mid-1970s, many students and young people who used to hang out in shops, bars, and restaurants in Shinjuku, in western Tokyo, moved on to neighborhoods farther out, includ-ing Shimokitazawa. This was the turning point for Shimokitazawa to become a "young people's neighborhood" (Kimura 2005).

In 1979, a music festival, Shimokitazawa Ongakusai, attracted more than 4,000 people. Small, live music clubs began to open in the neighborhood, and a bohemian culture developed. In 1982, the Honda Gekijo theater was established, and before long, small theaters began to open in the vicinity. In the late 1980s, the media began to promote Shimokitazawa as a place for the young generation, and this image of the neighborhood, whether it reflected the reality or not, emerged.

Shimokitazawa's uniqueness has been enhanced since then, mostly because other commercial districts in Tokyo have been bulldozed for urban redevelopment and lost the charm of pedestrian districts built to human scale. Moreover, with its many individually owned cafés, unusual fashion boutiques, and music outlets, Shimokitazawa is both politically liberal and hipster in style. According to the website of MTV Japan, "If such a thing as a core for Tokyo's independent musical and artistic culture can be said to exist, then the suburbs of Shimokitazawa and Koenji are where it's at" (Martin 2013). Online travel sites regularly describe Shimokitazawa as one of Tokyo's hippest neighborhoods.

Close Up: Azabu-Juban

Changes in the Retail Landscape, 1987–2013

Because of the subway opening and the redevelopment of Roppongi Hills, people can visit Azabu-Juban more easily than in the past. The increase in the "floating population" of visitors since 2000 has led to an increase in both the total number of stores and changes in the types of products they offer (see Table 1). The number of stores selling daily necessities to local residents, such as *tatami* (straw flooring mats) shops, pharmacies, and home appliance stores, has decreased. In contrast, the number of cafés, restaurants, and bars has rapidly grown, along with services like art galleries and fitness clubs. Generally new businesses illustrate the ABCs of gentrification: art galleries, boutiques, and cafés.

As on many local shopping streets around the world, the number of individually owned retail shops for everyday needs in Azabu-Juban has declined steadily since the early 2000s, while the presence of chain stores doubled between 2002 and 2013. This not only reflects the building of the subway station, it also indicates the general problems of economic competition and intergenerational continuity faced by individual store owners and their families. Today, individual store owners are forced to choose whether to continue their business or seek a different future for their children.

In Azabu-Juban, there are several Japanese sweets stores which have kept their businesses going for more than 100 years, even while experiencing the Great Kanto Earthquake and World War II. They have become a culturally valued symbol of Japanese tradition and local history as well. While continuing to make the same products in the same traditional way, they also try to develop new management and marketing strategies, such as online shopping.

Table 1 Retail Businesses in Azabu-Juban, 1987–2013

Business Type	1987		2002		2013	
	Total	No. of Chains	Total	No. of Chains	Total	No. of Chains
Clothing	32	0	33	0	43	3
Markets and Groceries	42	4	35	5	39	7
Cafés and Restaurants	72	1	78	4	114	8
Daily Necessities Other Than Food	52	1	55	1	37	6
Services	36	7	32	8	60	14
Others	19	0	27	0	28	0
Total	253	13	260	18	321	38

Source: Azabu-Juban Shotengai Sinko Kumiai (Azabu-Juban Shopping Street Promoting Association); *Juban Dayori* (monthly newsletter); the numbers of stores are calculated and compiled from the street map by Takashi Machimura and Sunmee Kim.

Nevertheless, for many individual shop owners, closing their business and choosing an alternative way of life seems unavoidable.

Since the 1990s, some older store owners in Azabu-Juban have closed their business, and either sold their house and land to developers or added more floors to the building that they own, becoming landlords rather than shopkeepers. In this case, they earn rent from the new shops that locate on the ground floor. Because a large number of buildings have been transformed in this way, the vertical scale of the entire shopping street is much higher than in the past.

Though many shops are new, they actively maintain the street's unusual fusion of "Japanese" and "cosmopolitan" character. In Azabu-Juban, you can easily find a shop selling doughnuts made of organic tofu, vegetarian cafés, and an upscale boutique with shirts for men, alongside a hundred-year old kimono shop and a typical Japanese *sembei* shop where the owner makes rice crackers in the front window of the store. Moreover, according to the food website www.bento.com, Azabu-Juban has a bar selling 13 Japanese craft beers, a restaurant where the Indian chef-owner creates South American dishes, and a Spanish restaurant whose Japanese chef spent nine years cooking in France and Spain. This is a combination of businesses that cannot be seen on most other shopping streets in Japan.

Nevertheless, although Azabu-Juban still has only a few chain stores, compared to other shopping streets in Tokyo, their number is continually increasing. Not only have McDonald's and Starbucks located in Azabu-Juban, but also branches of Japanese national chains in various types of businesses.

Can Family-Owned Shops Survive?

Currently the name "Azabu-Juban" is well known not only to Tokyoites but also to visitors from outside the city and tourists from overseas. Like Utrechtsestraat in Amsterdam, the atmosphere and shops are both chic and traditional. More pedestrians walk along the streets today than in the 1980s, when Tokyo, like many other central cities, lost population and faced decline. Seemingly, Azabu-Juban is a destination for consumers who are looking for a "distinctive" experience. Is this a good way for a traditional shopping street to survive?

Launching a mixed atmosphere that is both "Japanese" and "cosmopolitan" demands a complex strategy. In Azabu-Juban, it is supported partly by the unique character of the surrounding neighborhood, with more than a dozen foreign embassies and a fairly large, foreign residential population. Under the pressure of a globalizing economy, it would be easy to produce a clichéd response, simply leading to the reproduction of "cheap" mimicry all over the world. But Azabu-Juban adapted a "global" strategy in a more vernacular way, with the help of rich local resources and historical heritage. If Azabu-Juban is successful, it is because the street as a whole, and individual shops, have kept most of their authentic character.

Yet "authenticity" is a scarce cultural good that can easily be commercialized and lose its distinctive value. An "authentic" shopping street soon becomes a destination and, then, a target of investment. Particularly after the turn of the twenty-first century, this trend was exaggerated by the opening of the subway station and Roppongi Hills. Soon, Azabu-Juban, once considered a "hidden place," was discovered by the media. More stores opened, especially chains. Consumers came looking for the "real" and "authentic" Japanese culture, but with more chain stores opening, it was more difficult to find.

Gentrifiers also love Azabu-Juban. Its name has economic as well as symbolic value. In 2009, when a 38-story, luxury condominium building was constructed in a neighboring area of small, old houses, it was named City Tower Azabu-Juban, although its location is not really there.

Azabu-Juban teaches us about the ironic consequences of authenticity in this globalizing age. A commercially successful image of authenticity, even if it is just imagined, can cause the loss of the "original" authenticity. But perhaps this is a hasty conclusion. Shopkeepers are not only passive agents of structural forces like globalization, they are individuals who challenge, ignore, and appropriate them.

Shopkeepers' Stories

Kimono Art Sunaga, Azabu-Juban: Between Tradition and Innovation

Sunaga Tatsuo, who recently turned 70, is the fifth-generation owner of Kimono Art Sunaga, one of the oldest shops in Azabu-Juban. His store, located between a takeout sushi shop and a fresh juice shop, is one of the last surviving kimono shops in Azabu-Juban. It's a calm, neat space with various little fabric-covered objects, like wallets, notebooks, and card cases, in traditional Japanese patterns. Mr. Sunaga produces custom-made kimonos, but he also sells kimono accessories such as *obi* (the traditional wide sash), *maneki-neko* (the figurine of the beckoning cat) and children's *yukatas* (casual robes). Some of these are displayed on racks on the sidewalk in front of the store.

Inside, there is a small fitting room for measuring customers for a kimono, and many different fabrics are neatly piled on the shelves. The accessories and the fabric-covered objects are designed as a modern reinterpretation of Japanese colors and patterns, which makes Mr. Sunaga's store distinctive compared to a typical kimono store. Mr. Sunaga is proud that Hollywood celebrities such as Sarah Jessica Parker come to his shop to buy Japanese souvenirs when they visit Tokyo.

His ancestors, who came to modern Tokyo from the rural area of Gunma prefecture, started the family business in 1887. At that time Azabu-Juban was "a very prosperous area, because there was a rich residential area on the hills and in the side streets a licensed red-light district." When his father suddenly passed away, he was in his second year of university. "As the third son, actually I did not expect to take over the family business. But none of my siblings wanted to do it, so I become the fifth owner of the store, after my graduation in 1967." After two years of training in a vocational school for making kimonos, he genuinely engaged in the business:

> Now many of the stores have been turned into condominiums and people don't wear kimonos any more, but in the past there were many estates up in the hills. We used to visit the estates to sell luxury kimonos and people who lived there were our VIP customers. Those were the good customers! We let kabuki actors and celebrities run a tab, but they were bad at paying their bills. So times could be hard.

Yet the kimono store owner had to adapt to a changing culture. People don't wear kimonos as much as in the past, and textiles are, as elsewhere in the world, a prototype of a declining industry. He also faced a dramatic decrease in consumer spending at the beginning of the 1990s. Kimonos, after all, are expensive. A formal kimono can cost several thousand dollars. "At the end of the economic bubble we changed to the form of retail that's at our storefront today," Mr. Sunaga says.

"It was a risk, but I had to do something. I wanted to create a charming store that customers would want to visit. Our regular customers of the past have almost all passed away, and this industry is getting smaller and smaller. Nowadays there are few places to buy dry goods even in department stores." So Mr. Sunaga modernized his business strategies while emphasizing the traditional, Japanese image of a kimono store. This played well in Azabu-Juban, where shoppers are affluent and cosmopolitan.

As the chairman of the local merchant association, he is concerned not just with the survival of his own business, but with the future of the entire shopping street:

> The turning point of the local economy was the collapse of Japan's bubble economy. Quite a few of the store owners sold their plot of land when the price suddenly jumped in the bubble economy. They shut their business and moved to the suburbs. It was inevitable for them, I suppose. But since they opened the subway and there are more people walking down the street, the turnover of shops has been intense.

Mr. Sunaga sees the increasing number of studio apartments and "nightlife spots" as a sort of threat. His priority is to keep Azabu-Juban safe and quiet, elegant and upscale. "The candy stores and handmade chocolate shops are still doing well, as they do their business in the daytime. But as the places to drink alcohol increase, the less orderly it becomes."

His own survival strategy is to maintain the exquisite combination of tradition and innovation in the kimono shop. But survival means nothing to him if there is no neighborhood community:

> I don't want Azabu Juban to become like Roppongi, a tawdry, nighttime entertainment place. What I want is a town that gives relief, a town where you can walk with your family, a town in which I'd like to keep living. For example, there are many customers who come to visit the temple two or three times a year, saying, "When I come here, I feel like I'm coming home." I speak with those customers as old friends. "Oh, so you've come again this year." That kind of relationship is extremely important.

Blue and White, Azabu-Juban: Rediscovering "Japanese-ness"

You can guess from the name of Kate Yamada's shop, Blue and White, that she sells all kinds of traditional Japanese handicrafts. The traditional Japanese color scheme dominates the shop's décor, as well as many of the fabrics and ceramics that are on display. Ms. Yamada's interest in Japan began when she was a college student in the United States, studying Japanese history and culture. "I wanted to know a little more about Japanese society, so I came here to study. At that time I met my husband, and we got married."

In 1975, when she started the business with two other friends, she was a housewife with four children. The most important reason for choosing Azabu-Juban was simple; she lived there. But more than that, "it was also because at the time Azabu-Juban had the feel of an old Tokyo neighborhood." In the late 1970s, there were still arcades and a cinema which no longer exist, and the redevelopment of Roppongi Hills had not yet begun. "Local old ladies would often gather to talk on the roadside or in front of the greengrocers. It really was a lovely scene."

To Ms. Yamada, the most distinctive feature of Azabu-Juban is its diversity:

> I think that Azabu-Juban is the only place where old Japanese ladies live alongside diplomats and *yakuza* (gangsters). Japanese people always distinguish between themselves and foreigners, but I feel as though this is a place where I can live without thinking about such things. Foreigners enter into and frequent various shops so there is no cultural shock.

However, like Mr. Sunaga, Ms. Yamada says that the opening of the subway station and the redevelopment of Roppongi Hills brought huge changes. The traditional street scenes that she loved started, little by little, to disappear. In our interview, she openly declared herself against redevelopment:

> I think that Roppongi Hills did a lot of damage to this community. They only talk about the future, and the social networks of the past were all left unprotected. They destroyed the community and built a place of tall buildings, in other words a place without human communication. I think that they weren't aware of the importance of local community.

The opening of the subway station in 2003 was the biggest turning point. "I know it had some good effects on businesses, but I don't think that the local people wanted it. The atmosphere of the street changed a lot from what it was before."

This sudden modernization of Azabu-Juban's atmosphere contrasts with the products that she sells, which are almost all handmade by Japanese artisans. "I enjoy pieces of work that use natural materials, are colored beautifully, are made by people who studied the ancient traditional designs as best they can, and yet with a slightly modern feel."

Her main customers at first were foreign residents and tourists, rather than Japanese:

> When I started this store, Japanese people didn't consider their own culture to be particularly valuable at that time. So I was worried that traditional objects were being forgotten and falling out of use. I wanted to make use of the traditional crafts a little more in everyday life.

But times have changed for the handicrafts shop as for the kimono shop: "Now lots of my customers are Japanese." The "authentic" combination of Japanese and Western that made her store attractive to foreigners is now attracting Japanese customers who are looking for "tradition." Although Mr. Sunaga has modernized the kimono shop, Ms. Yamada must emphasize tradition. Both continue to do well in Azabu-Juban, but the street is changing in ways that they do not altogether welcome.

Less noticed but no less important, Azabu-Juban continues to welcome migrants. In earlier years, they came from rural Japan. They learned trades like dressmaking in Tokyo, opened boutiques in Azabu-Juban, and became successful. Today, a migrant may come from Sri Lanka, learn the Japanese language, get a job as a waiter in a French restaurant, and rise to become the manager. Migrants continue to provide the goods and services that nurture Azabu-Juban's upscale, cosmopolitan character, becoming integrated into—and sustaining—the cultural ecosystem of the street.

Shimokitazawa

Changes in the Retail Landscape

Shimokitazawa is filled with more than a thousand shops, restaurants, and service businesses (see Table 2). This is not only a much greater number of stores than in most shopping districts of Tokyo—three times as many as in Azabu-Juban—the stores are much more densely packed into the area.

Eating and drinking places predominate: Shimokitazawa has 283 restaurants and 140 cafés and bars. But there are almost 400 "services" of various kinds and more than 200 clothing stores. However, there are not so many places to buy groceries and food: only three supermarkets and seven convenience stores.

Table 2   Retail Businesses in Shimokitazawa, 2011

Business Type	Total	Individually Owned Stores	Regional Chain Stores	National Chain Stores	Unidentified
Restaurant	283	175	81	25	2
Café or Bar	144	129	12	3	0
Sale of Goods	176	102	58	15	1
Grocery and Food	20	15	5	0	0
Clothing	218	139	70	7	2
Convenience Store	7	0	0	7	0
Supermarket	3	0	1	2	0
Services	376	261	81	34	0
Total	1227	821	308	93	5

Source: Walking census, Keiro Hattori, and students, Meijigakuin University.

In this neighborhood, the "services" include 101 hair salons and spas, 50 medical offices, 38 real estate agents, and 20 massage parlors. From this retail landscape Shimokitazawa appears to be a "lifestyle" shopping district oriented toward young consumers.

The membership lists of the four local merchant associations in Shimokitazawa show that the types of stores have not changed very much since the 1990s. However, although real estate businesses are decreasing, a variety of "other services" are increasing, mostly hair salons.

Moreover, the turnover of shops in Shimokitazawa is quite high. In just two years, between 2011 and 2013, 433 new shops opened, a number equivalent to 33 percent of all the shops there. Although some shops survive for many years, many cannot.

In Shimokitazawa as in Azabu-Juban, the majority of businesses, 67 percent, are small and individually owned. Most of the restaurants and bars are individually owned, and the grocery and clothing shops as well. Chains, however, dominate the few supermarkets and convenience stores.

In contrast to Azabu-Juban, which has one merchants' association, Shimokitazawa, because of its size, has four. But they have seen a steep decline in their membership since 1990—not because there are fewer stores, but because fewer store owners want to join.

This is mainly due to three reasons. First, the increase in chain stores decreases the number of potential members, because chains hire a manager with limited autonomy in decision-making. Even if managers of retail chains want to join a merchants' association, they cannot take responsibility for the policies of their shop. Second, the benefit of joining a merchants' association has decreased because more business owners rely on the Internet for information and customers. Third, the strength of community on shopping streets has been decreasing, allowing many shop owners to get away with "free-riding."

The four geographical quadrants of Shimokitazawa have specialized concentrations of businesses. Minamiguchi, to the south of the station, has many second-hand clothing stores, each with its own unique character, and many music clubs and music bars. Because of these businesses, this area attracts more young customers than the others. Azuma-kai, to the west, has a lot of small Japanese-style eating places (koryori-ya) and drinking places, as well as small theaters. Kitaguchi, to the north, has a lot of stores that sell food and everyday goods, but recently trendy restaurants and new boutiques have started to change its character. Ichibangai, the oldest shopping area in Shimokitazawa, has some long-established shoe stores, liquor stores, and restaurants. However, new second-hand clothing shops are beginning to open there as well.

Dangers of "Authenticity"

Unlike Azabu-Juban, which has a long and distinctive history, a hundred years ago Shimokitazawa was just an ordinary farming village. Before World War II, with the building of the railroad and new urban housing, it developed into a typical Tokyo suburb. The war helped to create a new character for Shimokitazawa for two reasons. First, the good fortune of not being bombed helped to sustain the old physical structure of pedestrian-friendly, narrow streets and alleys. Second, the opening of a black market near the railroad station made the area attractive as a shopping destination.

However, these were not enough to create an "authentic" character. The authenticity of present-day Shimokitazawa comes from the hippie culture that was imported mainly from the United States in the late 1960s and early 1970s. The shops and clubs in Shimokitazawa transmitted this culture to Japanese consumers through jazz and rock music, fashions, alcoholic beverages, coffee, books, and other cultural products.

Between the 1970s and 1980s, a "new" authentic culture emerged in Shimokitazawa that blended new Western trends with traditional Japanese culture. Rock musicians based in Shimokitazawa, such as RC Succession and Jagatara, began to write and sing songs in Japanese that no longer just imitated American or British rock bands. Some curry restaurants in Shimokitazawa began to create new recipes that you cannot find in India. In addition, the opening of the Honda Theater in 1982 helped the neighborhood to become a mecca of amateur theatrical activities. Consequently, many small theaters as well as music clubs sprang up in the neighborhood. These changes allowed Shimokitazawa to become an incubator for new urban, do-it-yourself culture similar to the East Village in Manhattan, and Williamsburg in Brooklyn.

For the first time, the 2013 Michelin Guide to Japan designated Shimokitazawa a one-star tourist destination. The guidebook explains that the criteria for this include "richness of cultural assets, plenty of leisure activities, and *authentic* charms" (emphasis added). This shows that, like Azabu Juban, Shimokitazawa has been able to create a strong sense of local authenticity in a global culture and global economy.

Unlike Azabu-Juban, Shimokitazawa has not done this by enhancing traditional "Japanesque" values but by modifying Western culture through Japanese filters. Today's Shimokitazawa began as a place for the display of American subculture, but gradually it began to create a unique Japanese culture.

Ironically, this success has had negative effects on the shops, clubs, and bars. Shimokitazawa's image as a hip shopping and entertainment district has raised the rents. Some local shopkeepers, who have been the main actors in creating Shimokitazawa's authenticity, are having a hard time making ends meet.

Moreover, the government's long-standing plan to construct a wide road on the northern side of Shimokitazawa is supported by landowners. According to the plan, the big road will permit buildings in the district to have a higher floor-area ratio, which means that the current buildings can be made taller, or demolished and replaced by high-rises. The new road and buildings will likely destroy Shimokitazawa's pedestrian-friendly physical environment. This in turn risks destroying the "living" shopping street's authentic social and cultural character.

Business owners, especially all those who have created Shimokitazawa's post-1970s authenticity, are well aware of the risks they face.

Shopkeepers' Stories

Mother, Shimokitazawa: Creating "New" Authenticity

The rock bar Mother stands on the southern edge of Shimokitazawa, roughly a five-minute walk from the railroad station. According to *Time Out Tokyo*, the décor "resembles a mix of gingerbread house, treehouse and pub," but Mother "is in fact a classy establishment" (*Time Out Tokyo* undated).

Chizuko Yamazaki, the bar owner, opened it in 1972, at the age of 24. She did not have a steady job but did part-time work to survive and, in her words, "was wandering." Because she was involved in the student protest movement, she decided she couldn't work in a normal company. Ms. Yamazaki says she had no desire to make a profit: "I just wanted to make a living. I thought if I owned a restaurant, I would at least have something to eat."

She came to Shimokitazawa by chance. She liked the neighborhood, and found a good space to rent. But she needed 500,000 yen (about $5,000), which she was able to borrow from her mother. "I did not have any savings at the time, but my mother was willing to invest in me." Ms. Yamazaki named the bar Mother to show her appreciation, and also because she liked John Lennon's 1970 song with the same title.

At first, she and her customers were the same age. She was not very polite, hardly ever greeting them with the traditional welcoming phrase, "*Irrashaimase*." But she had fun drinking with customers, and found rock music very exciting. Customers would bring records for her to play at the bar, by such bands as King Crimson and the Allman Brothers. "Many customers were students. However, they did not seem to be so poor. They taught me about music."

Bars like hers were not so common in those days, even in Shimokitazawa, but there were a few, and customers used to make the rounds among them:

> At the time, we all felt Shimokitazawa was like a village. When the bar was closed, I used to go out drinking in other bars in Shimokitazawa. We talked a lot about movies and plays, and sometimes about politics. Many people related to the businesses that started to open in Shimokitazawa at the time. The difference between those days and today is that young

people now are more sophisticated. Back in the day, young people drank to get drunk. In fact, I drank too much when I was young. I worked from 2 in the afternoon to 4 in the morning, and slept from 7 a.m. to 12 noon.

At the time, I already had a strong affection toward Shimokitazawa. It was like my backyard. I liked the village feeling. And the shopping was quite convenient. I did not need to take a train to go shopping, and I did not even need to ride a bicycle. It was so convenient.

I moved to Shimokitazawa when I opened the bar. All the young people also lived around Shimokitazawa.

My customers, who were college students when they first came to Mother, grew up. They began to make movies and play in bands. One of those customers started a band which later became Jagatara, a legendary rock band of the 90s. The bar functioned like an incubator. It was a bit like a training experience for young customers.

Customers did not choose the bar; we chose the customers.

But recently business has not been good:

The disaster of the Fukushima nuclear power plant has been quite damaging for our bar and also for Shimokitazawa. It has already been two years, but I cannot see that things are getting better. Many stores in the neighborhood are being taken over by cabaret-clubs or cell phone stores.

The plan to build the new road also poses risks:

It will totally change the landscape. The road will enable developers to construct high-rise buildings. This small, village-like Shimokitazawa does not need to have tall buildings like Shibuya or Roppongi. It will threaten our survival. Shimokitazawa is a walking environment. We do not need a 26 meter-wide road.

I opened a bar in Shimokitazawa. I also live here and raised my children here. I am going to fade away before long, but I strongly feel that I need to preserve this neighborhood for my daughter and for her daughter. The neighborhood is where you live. You don't let go of a place where you live so easily because next generations will also live there.

When Ms. Yamazaki opened the bar in 1972, she did not imagine that she would still be there decades later. Now her grown-up daughter also works at the bar, hoping that one day she will take over the business.

Never-Never Land, Shimokitazawa: Opposing Development

Kenji Shimodaira is the third owner of the Never-Never Land bar, which, despite its rather isolated location on the edge of the northern shopping district in Shimokitazawa, has been open continuously for more than 35 years.

Colorful and hip, exotic but not very sanitary, the bar really has had several lives. It began on the next block, where the first owner, Mr. Masai, created a bar in the small apartment that he was subletting without the owner's permission. He named the bar Them, after a 1960s Irish rock band led by Van Morrison. He also was a music promoter, inviting big-name performers, including the Blues Brothers, to concerts in Japan. Mr. Masai managed his bar for six years but eventually was pushed out by the apartment owner.

Mr. Matsuzaki, who was a professional photographer, talked the apartment owner into letting him rent the place and continue using it as a bar. He called his bar Never-Never Land, from Peter Pan, saying that it was a place for adults who never want to grow up. Rents were cheap then in Shimokitazawa, making it easy financially to create a place where people could gather together and enjoy hanging out.

The current owner, Kenji Shimodaira, visited the bar for the first time when he was eighteen. He was attracted by the big sign that still says "ROCK BAR." Born and raised in the mountainous region of Iida City in Nagano Prefecture, Mr. Shimodaira had come to Tokyo in 1980 to attend dental school. He decided to rent an apartment in Shimokitazawa because an acquaintance of his mother lived there. A "rock and roll teenager," as he says, he was excited to be living in Tokyo, and wanted to explore the city.

At that time, Shimokitazawa was not the hipster district that it later became, but there was already a trend of "cultural people," especially musicians, coming there. There was a traditional Japanese restaurant called Kagetsu, where many big names of Japanese and worldwide music and other celebrities hung out, like the Rolling Stones, Louis Armstrong, Junko Koshino, Kenji Sawada, Yosuke Yamashita, and Yumi Arai. Never-Never Land itself played rock music, and celebrities including the Japanese musician Goro Nakagawa and Tunk (Mitsuo Terada) visited.

"This neighborhood [has always liked] rock music," Mr. Shimodaira says. "You can tell by the names of the bars: Mother, Berlin, Gasoline Alley, Trouble Peach. These names are taken from the titles of rock albums." He repeats what Chizuko Yamazaki, the owner of Mother says. "The neighborhood chooses the people and the customers, here. It is our way or the highway."

Despite the "ROCK BAR" sign, when Mr. Shimodaira first entered the bar, he saw two pairs of guests playing *shogi* (Japanese chess). He happened to be a very good Japanese chess player; in fact, he was the champion of his neighborhood when he was growing up. So he played Japanese chess with the other customers, and everyone admired his skill. From that day, he was accepted as a "regular" customer of the bar.

However, to get respect in Shimokitazawa in those days, you needed to be an artist, someone who has an "aura" or is recognized as being inspired. Mr. Shimodaira explained:

> I was very strong at Japanese chess. I even became a Japanese backgammon champion. But I was not an artist. I played the guitar and loved rock music, but I was not good enough. I did not have the talent to get respect in this neighborhood. So, it was not comfortable for me, being in Never-Never Land.

Indeed, Mr. Shimodaira was not a popular figure. His girlfriends were often stolen by more attractive men in the neighborhood. The bar owner, Mr. Matsuzaki, told him that he had no charisma. Kenji Shimodaira is sure that Mr. Matsuzaki never really approved of him.

Yet, when Matsuzaki remarried in 2000, his new wife Kyoko said, "There is something in Kenji." And when Matsuzaki passed away in 2003, Kyoko asked Mr. Shimodaira for help so she could keep the bar open. But operating the bar was very difficult financially. After two years, Kyoko began to think of closing down.

At that time, another rock bar in Shimokitazawa called Gaja shut its doors. People in the neighborhood, including Mr. Shimodaira, met to discuss how to reopen it. An assemblyman of Setagaya Ward asked him for help. Kenji Kaneko, who later teamed up with him to form the Save Shimokitazawa association, which would fight against the plan to widen the road, also joined the meeting. Although the participants could do nothing to resurrect Gaja, they eventually became the core members of Save Shimokitazawa.

The loss of Gaja really hurt Mr. Shimodaira. He thought that if Mr. Matsuzaki's widow closed Never-Never Land, it would be quite devastating not only for him, but for Shimokitazawa as well. Yet, the monthly rent was 240,000 yen (about $2,000), and the economy was in a freefall.

Mr. Shimodaira knew that if he took over the bar's ownership, it would cost him a lot of money to keep it open. But he also knew that if he worked hard as a dentist, he might be able to afford it. So in 2006, he bought Never-Never Land. "I probably love Shimokitazawa," he says:

> I was never a hero. But I found my wife here. I felt like I owed something to Shimokitazawa. But few people in the district respected me. I thought I should become a more important figure. I wanted to show what I could do. This neighborhood lacked a "producer." I thought I was the one who could act as a producer.

However, the shopping district needed more than a "producer." In addition to high rents and the long-standing plan to widen the road on the north side of Shimokiazawa station, in the 1990s the privately-owned Odakyu Railway Line proposed building new tracks through the area. Public discussion focused on whether the new tracks would be built underground or overhead, in a viaduct. Many artists opposed the viaduct. To give voice to their protest, several musicians organized a live show, the "Shimokitazawa Revolution," at the Kitazawa Town Hall. The well-known performers included Mikijiro Taira, Minako Yoshida, and Kenji Sawada.

Mr. Shimodaira found his niche in the opposition to these projects. He became a leader of the Save Shimokitazawa movement (Littler 2011).

"If the road was built, Trouble Peach would be torn down. I could not take that." When community opposition delayed both plans, for the road widening and the railway viaduct, for years, and the viaduct was replaced by a tunnel, many people in Shimokitazawa began to look up to him.

Despite his success as a community activist, owning the bar has not been easy. Mr. Shimodaira confesses that he lost more than 3.5 million yen ($35,000) on the bar last year. But, he has no intention of giving it up. "As the owner of Never-Never Land, I would like to show young people the essence of Shimokitazawa," he says.

He recalls that the writer who described the bar in the Michelin Guide told him that it represents the neighborhood's authentic identity:

> So I try to make this bar like the neighborhood's living room. After work, you come to your living room to relax. People come to Never-Never Land when something good happens, and also when something sad happens. I want to make this bar an icon of Shimokitazawa's culture.

Both Never-Never Land and Mother act as cultural incubators and represent authentic local culture. Their owners gladly take on this role because they love the neighborhood. Yet like many bar and restaurant owners in Shimokitazawa, neither of them makes a profit.

Rents in Shimokitazawa are rising because the shopping and entertainment district is so popular. If individual owners cannot pay higher rents, chain stores are ready to move in. The shopping streets to the south of the railroad station have already been taken over by national and transnational chains like McDonald's, KFC, Docomo, Yarukijaya, Goemon, and Thirty One Ice Cream.

This is an ironic result of Shimokitazawa's authentic appeal. Visitors come because of the unique small shops and restaurants. This draws the attention of the chains. But their deep pockets cause landlords to raise the rent, wiping out the very businesses that are the source of Shimokitazawa's success. The big road construction is also hurting the small shops and bars. The ecosystem of narrow shopping streets and alleys may not survive these blows.

Facing the Future

Like every other shopping street in the world, the small business owners of Azabu-Juban and Shimokitazawa confront a keen and often unfair competition with huge chain stores, mega-malls and outlets, and global retailers. As individuals and as local shopping districts, they have developed different strategies to meet the threat.

But Azabu-Juban and Shimokitazawa share an impressive tolerance towards different cultures and tastes. This tolerance creates an authentic public space in the shopping street, which contrasts with the privatization that has become common around the world. While "authenticity" appears to be a very attractive business strategy for these two districts, it needs to do more than represent the past. An "authentic" local shopping street must restate new and multiple cultural narratives, based on changing transnational migrations and adaptations.

Agents of globalization know authenticity's true value. They try to promote "surviving" or even "resisting" streets to the global list of shopping destinations. Azabu-Juban and Shimokitazawa, featured in Time Out, the Michelin Guide, and shopping and entertainment websites, are two successful cases. However, their ability to remain "living" shopping streets depends on whether they are given the opportunity—by developers and government—to remain spontaneous and on a human scale.

Acknowledgements

This chapter is based on interviews, land use observations, analysis of local documents, and a walking census of stores carried out from 2010 to 2013 by Sunmee Kim and Takashi Machimura (Azabu-Juban), and Keiro Hattori and his students (Shimokitazawa). We express our gratitude and appreciation to all the business owners, managers, and employees who spent time with us in the streets, especially Chizuko Yamazaki and Kenji Shimodaira.

References

Kimura, Kazuho. 2005. "Wakamono no Machi Shimokitazawa no Tanjou." In *Aruku Tanoshisa no aru Machi, Shimokitazawa no Saihakken*, Shimokitazawa Symposium. Kitazawa Town Hall, Setagaya, Tokyo, July 16.

Liebs, Chester. 2011. *Sekai ga shousanshita Nihon no Machi no Himitsu [The Secret of Japanese Cities that the World Admired]*, translated by Keiro Hattori. Tokyo: Yosensha.

Littler, Julian. 2011. "Staving Off the Bulldozers for One More Round." *Japan Times* (October 14). Available at www.japantimes.co.jp/life/2011/10/14/food/staving-off-the-bulldozers-for-one-more-round/#.UokUGeKE7HE (accessed November 17, 2013).

Martin, Ian. 2013. "Scenes: Shimokitazawa and Koenji's Eerie Backstreets and Indies." Available at www.mtv81.com/features/specials/tokyo-scenes-koenji-shimokita (accessed November 13, 2013).

Time Out Tokyo. Undated. "Mother." Available at http://www.timeout.jp/en/tokyo/venue/402/Mother (accessed November 17, 2013).

Tokyo Metropolitan Government. 2001. *Census of Commerce 1997: Tokyo Metropolis Report.* Available at www.toukei.metro.tokyo.jp/syougyou/1997/tob6f700.htm (accessed January 15, 2014).

Tokyo Metropolitan Government. 2009. *Census of Commerce 2007: Tokyo Metropolis Report.* Available at www.toukei.metro.tokyo.jp/syougyou/2007/sg07v20000.htm (accessed January 15, 2014).

Tosa, Shigeki. 2011. "Hajime Matsumoto: Influence Politics with Street Demonstrations." *Asahi Shimbun* (July 27). Available at http://ajw.asahi.com/article/0311disaster/opinion/AJ201107275009 (accessed November 16, 2013).

Yoshimoto, Banana. 2010. *Moshi Moshi Shimokitazawa.* Tokyo: Mainichi Shinbunsha.

CHAPTER 8

Local Shops, Global Streets

PHILIP KASINITZ, SHARON ZUKIN, AND XIANGMING CHEN

By this point it should come as no surprise that the authors of this book *like* local shopping streets. We believe—and we hope readers are now convinced— that the streets we have written about play an important role in shaping how social life works in today's global cities. We like the civility, and yes, the cultural diversity of local shopping streets. They are what men and women picture, dream of, and move to, when they think of the "authentic" city: an experience of living in a dense, socially diverse, sensually stimulating environment.

We also like the impact of local shopping streets on creating a walkable, bike-able, environmentally sustainable city. We feel that such streets have value, and that preserving them is an appropriate task for urban policy.

But "preserve" in which way? Should the city government maintain the stores, people, or physical scale of the space? And which policy tools would work best: planning, zoning, rent regulation, or deliberate recruitment and exclusion of specific shops? Let's not forget that local shopping streets are, at their core, urban marketplaces, which require continual realignment between the needs and wants of multiple buyers and sellers. The interests of building owners (landlords) must be balanced against those of store owners (tenants), and the tastes of affluent shoppers and hipsters ("gentrifiers") must be balanced against those of longtime residents with more modest means and less overtly cultural ambitions.

Complexity and Sociability

It is clear that local shopping streets are complex urban spaces. Though they are primarily organized around commerce, they are as valuable for their extra-economic functions as for their contributions to the local economy. Local shopping streets do generate jobs and taxes, though probably not more than other forms of retailing. But that is only one part of their social value.

Small-scale shopping streets help to create a sense of place. They form a visual face of local identity, a node of distinctiveness in an increasingly standardized city. The more they maintain their distinctiveness, the greater the city's competitive advantage for shoppers, residents, and tourists from around the world.

Local shopping streets are also a key site of urban sociability and interaction. At the least, they support a co-presence of strangers that is more intimate than in a subway car or a public park but less direct and certainly more enjoyable than in the offices of a municipal bureaucracy.

We cannot emphasize enough that local shopping streets are spaces of everyday diversity. In a world of perpetual journeys and migrations, they are often the first meeting place between people from different parts of the globe who are brought together by rituals of commerce rather than by shared cultural rituals. At their best, the super-diversity of many local shopping streets eases the way towards civility and tolerance as normal conditions of urban public life. In a low-key, practical way, these streets can create a "corner-shop cosmopolitanism" in which everyone feels at home (Hall 2012; Wessendorf 2014).

Moreover, small, individually owned restaurants and shops remain a significant way of integrating new immigrants into the mainstream economy and culture. Not only in New York, but in the U.S. today, the vast majority of small retail business owners are foreign born (Kallick 2015). Immigrant-owned shops create significant economic, social, and cultural benefits for all.

Risks of Globalization and Gentrification

But globalization poses great risks to most of the local shopping streets in this book. They confront both the need to integrate strangers into public life without fear or bias, and the need to support economically and socially marginal groups without holding out an unrealistic lifeline to obsolete businesses and trades. We may mourn the demise of record stores or the disappearance of a favorite diner, but we understand the efficiencies of modern, computerized, large-scale retail operations. We see a need to seek a balance between old and new businesses, and large and small shops, with support for individual and local ownership.

Another issue is gentrification. Jane Jacobs (1961) warned years ago of the dangers of "cataclysmic" urban renewal. Today, we must be alert to the risks of equally cataclysmic gentrification (Zukin 2010).

Commercial gentrification brings both new life and new problems to local shopping streets. Even a small amount of capital investment can bring a shiny new look to a shabby shopping street. Good write-ups in the media and favorable reviews bring more new investors: shoppers, diners, residents, real estate developers, and more new retail entrepreneurs. But the growing appeal of the street, and of the neighborhood around it, is bound to raise rents.

We have seen what happens next. Longtime residents may be priced out and cast culturally adrift by the global ABCs of gentrification—art galleries, boutiques, and cafés—while building owners, local government, and nearby residents are pleased by the neighborhood's "rising" image. Gentrifying businesses draw attention, and sometimes an international reputation as a "hip" and "trendy" site. This promotes the street as a whole and "brands" the surrounding residential neighborhood. Property values rise, which is good for homeowners, landlords, and city tax revenues. But sooner or later, higher rents will displace lower-price and lower-volume stores.

In the increasingly anonymous global city, the local shopping street offers a kind of "home" to urbanites. However, this sense of "home" is threatened by rapid gentrification and globalization. Change is inevitable, but the speed of change in recent years, and the inability to cushion the loss of homes and familiar shops, has been catastrophic.

Yet as Jan Willem Duyvendak (2011) points out, "home" has always been a tricky concept. It is important for people to feel they have a home in the city, Duyvendak notes, especially groups who are excluded elsewhere. But a group that establishes a concentration of retail stores and services, such as gays in the Castro in San Francisco, may push out another group, such as working class whites. Don't all groups have a right to create a home in the city, a place where they feel safe? The challenge is to balance a sense of "moral ownership" among different groups, even though some may have greater resources of financial and symbolic capital.

We recognize that not all local shopping streets can, or should be, preserved like flies in amber. Nor does it make sense in the modern world for all commercial activities to be conducted on a small, intimate scale. Indeed, the argument for small stores and local shopping can easily turn into a form of elitism. Small stores lack economies of scale, leading them to charge higher prices than large retail chains and discount stores. "Interesting" shops and "artisanal" commerce are often a luxury for consumers who don't have to worry about cost. Trying to ban big retail chains and fast-food franchises does not serve the interests of those who need low prices to feed and clothe their families.

At the same time, local shops contribute meaning to the life of the city and its residents, including, or maybe especially, for low-income and socially marginalized residents. This brings us to a dilemma. Low-income communities need low-price stores, which are often branches of national and transnational chains. Yet all communities benefit from the feelings of moral ownership that "their" local shops can provide. Aren't local shops a way of asserting a right to the city?

As is so often the case, the key here is balance. Can we guide the ecosystem of local shopping streets instead of allowing it to spiral out of control?

Role of the Local State

It is worth looking back at the streets presented in this book to see that in none of them does the social and cultural ecosystem emerge from the vision of urban planners or state bureaucrats. While cities from New York to Shanghai, and their nation-states, represent vastly different systems of urban governance, one of the striking similarities between them is the limits on the role planners can play. Even in the most centrally planned system in our study—the People's Republic of China—both of the shopping streets we looked at have benefitted from the state's selective *inattention*, and sometimes *inaction*, at crucial moments. Allowing private owners of apartments to create commercial out of residential spaces, and allowing rural migrants to set up small stores in these spaces, serves many different social needs, all without official approval.

There are also real differences between cities in the role that the state is expected to play in urban redevelopment. In Amsterdam and Berlin, when the dominant groups in society see a street as a "problem," local planning agencies take it as their mission to intervene with a degree of micro-management that would be unthinkable in New York, Toronto or Tokyo. At the same time, resistance from local shopkeepers, as in Tokyo, may limit planners even when law does not.

Finally, different actors—namely, shopkeepers, longtime residents, and urban planners—may have quite different views about what makes a street a "problem." In Europe and North America, shopping streets catering to transnational migrants from the Global South stir more anxiety than those catering to migrants from the Global North. Internet cafés in heavily immigrant shopping streets are regarded with more suspicion than in streets where migrants are few. Gambling casinos in Berlin, cell phone stores in Amsterdam, and check-cashing stores in Toronto are kept under police surveillance. In these and other cities, immigrants' shopping streets are seen as disorderly and problematic by planners and often by the native-born population. Yet many local residents and visitors see the same streets and businesses as a hub of social life—*their* social life.

Global Toolkit of Urban Revitalization

Though good local shopping streets do not have to be planned, none of the "revitalized" streets we observed results from "pure" market forces. From New York to Shanghai, the same image of a successful shopping street prevails, part of a global toolkit of urban revitalization. And it looks very much the same all over the world: "curated" boutiques, restaurants, cafés, and bars, "gentrification by hipster," or, at any rate, gentrification. This model did not just spring forth from the sidewalk; it is heavily promoted by private investors and public officials throughout the Global North and, because of their influence and funding, and the international organizations that they form, it has spread through the rest of the world as well (Robinson 2011).

Though public officials in different cities take different approaches, reflecting local institutional histories and cultural expectations, the state is never absent. City government—the local state—creates a baseline by building infrastructure and keeping it in good repair (or not), establishing policing strategies for public safety, mandating land uses through zoning laws, and keeping track of business practices, from enforcing health and building codes to labor and immigration laws. At times, as in New York and Toronto and an increasing number of cities, the local state cedes authority for managing the public space of a shopping street to a public–private institution such as New York's business improvement districts (BIDs) or Toronto's business improvement areas (BIAs). This step eases pressure on the public budget, but does nothing in the streets that we studied to help individual businesses survive, especially the smaller ones. Wherever we went, local shopkeepers told us that the business improvement group on their street did nothing to address their most pressing concerns. Moreover, BIDs and BIAs often work to gentrify a street, not to promote a balance between low-price and "curated" shops. This is especially true when BIDs or BIAs represent the interests of building owners rather than store owners.

Economic Vitality and Ethnic Diversity

An area in which policies towards local shopping streets show different *national* political priorities is in the integration of new populations. This is an important issue because shopkeepers are so often newcomers. In China today, and historically in North America, Europe, and Japan, the majority of shopkeepers have been rural-to-urban migrants. Increasingly, however, they are transnational migrants who look different from the native-born majority and speak different languages. In many places, the aesthetic diversity represented by local shopping streets is celebrated, particularly when it is contrasted to the "sameness" of global chains. Yet social and ethnic diversity has been more difficult to accept. As Steven Vertovec (2012) notes, "diversity" holds very different meanings in different political contexts.

In the North American cities that we studied, ethnically specific shopping streets are often praised. "Chinatowns" and "Little Italys" were once seen as undesirable ghettos, but today, they are urban amenities and tourist attractions. City governments try to preserve the ethnic character of the commercial activity, sometimes long after the ethnic residential population has moved on (Anderson 1995; Aytar and Rath 2012; Kosta 2014). This fits well with North American ideas about ethnic pluralism. Even if these shopping streets are ethnically homogenous, their patchwork presence is seen as making the city as a whole diverse, a "gorgeous mosaic" in the words of former New York City mayor David Dinkins.

In Western Europe, by contrast, such ethnically identified shopping streets are often seen as a threat to social integration. A strong "ethnic" or "immigrant" presence in Europe is feared as a sign of fragmentation or even "ghettoization," which the state feels responsible to prevent. Yet the state's effort to "diversify" the street and reduce the concentration of similar stores, as happened on Amsterdam's Javastraat, ends up displacing immigrant store owners and replacing them with more native Dutch merchants. Managing "diversity" in cities then becomes a way of installing, or re-installing, a traditionally dominant majority. Yet in both North America and Europe, planners' ideas are often far behind the super-diversity that is developing on more shopping streets all the time.

Asian cities provide different models of integrating newcomers. Both Shanghai and Tokyo have low rates of international immigration. For this reason, little public attention, certainly very little negative attention, is paid to its conflictual politics. At the same time, Shanghai is home to a large and growing number of expats from the Global North. These well-educated, relatively affluent foreign nationals are thought of (and think of themselves) as very long-term visitors rather than as traditional immigrants. Yet they have a huge impact on the central city where they tend to live and work.

Though Tianzifang is officially designated a cultural industry zone, it has really been transformed into a global consumption space. Foreigners are welcome to set up local shops there, and English is heard almost as much as Mandarin. Native Shanghainese, we might point out, is heard less often in these central consumption spaces than on the periphery. At the same time, internal migration from other parts of China is also rapidly diversifying the city, despite strict national government attempts to control it. With so many internal migrants looking for jobs, the local government tolerates them opening and working in stores on streets such as Minxinglu, without legal permits. This absorbs unemployed rural migrants into the urban economy and helps maintain social stability, which concerns the local government a great deal.

Even if they operate without the required licenses, local shops create public spaces that, if not "ethnically" distinct in the ways we see in North America and Europe, do produce more socially and culturally diverse streets than we find in other areas of the city, where urban shopping malls abound. Though Tianzifang offers the global ABCs of gentrification—art galleries, bars, and cafés, many owned by foreigners and overseas Chinese—Minxinglu almost entirely hosts ordinary businesses owned by internal migrants, including a halal restaurant owned by migrants from western China. On such streets people who literally have no legal right to the city have begun to stake out their social rights.

In Tokyo, too, immigrants are not highly visible on most neighborhood shopping streets. Indeed, these streets often display a conspicuous sense of "Japanese-ness" in contrast to the global consumer identity seen in chain stores and shopping centers. Here traditional national cultural identity is both a prod-uct—kimonos, rice crackers, special pastries—and a form of promotion for the street.

In upscale Azabu-Juban, the few non-Japanese employees working in restau-rants and shops are learning "Japanese" ways of running a business. Some plan to transfer this knowledge back to their home country if they return there. Japanese-ness, not ethnic diversity, is on sale, regardless of the ethnic and national origins of shopkeepers and restaurateurs. In a different type of shop-ping district, Shimokitazawa, hipster bars and vintage clothing boutiques that would be at home in any city of the Global North establish a different "brand" of Japanese-ness. Though these businesses represent aesthetic and generational diversity—because they mostly cater to young people with specific tastes—they produce a distinctive local identity.

Need for Public Policies

As these examples show, neighborhood shopping streets are both extremely global and intensely local. They are increasingly ethnically diverse. Though they are structured by both local and national political priorities, they are also shaped by a global toolkit of strategies of urban revitalization. These strategies consciously aim to promote local distinctiveness, but perversely run a risk of making all local shopping streets more alike.

Local shopping streets have many "authors," for they are created by a combi-nation of cultural, market, and state regulatory forces. If they are too microscopically planned or repressively regulated by the state, they will lose their economic value to merchants and lose the cultural value that attracts customers. Today, however, the threat more often comes from the market. Commercial gentrification and globalized consumer culture destroy traditional shopping streets in ways that might be economically efficient (at least for large chain stores) but ignore the extra-economic benefits small, locally owned shops provide.

How, then, might political action, whether filtered through the state or aris-ing from social movements like "Buy Local" or "Slow Food" campaigns, help to stabilize these streets without destroying their capacity for change? How can we use the state and public–private organizations like community development corporations and BIDs to restore a balance between large and small retail, between the global and the local? How can we achieve a balance between massive change that uproots small store owners and the coherence and longevity of a shared, moral ownership?

In many places public policies have begun to address the need to support local shopping streets. These policies seek to protect cities from standardization by large chains, on the one hand, and encourage small-scale and individual ownership, on the other. Diversified shopping streets, advocates argue, will produce a lively neighborhood, not necessarily in terms of social and ethnic diversity, but in terms of physical fabric, with many different kinds of shops to delight the senses, and doors and windows open to the public space of the street. Diversified shopping streets also sustain a socially functional temporality, bringing people out to the street twenty-four hours a day, seven days a week, with no "dead" times when the street is empty and frightening. Finally, diversified shopping streets create a sense of comfort, providing an expansive place between their homes and the world for shoppers and local residents.

This is not an easy agenda, given the lure of big-box stores with their huge parking lots on the outskirts of cities, as well as the attraction and convenience of shopping online and on mobile devices. Yet we see such efforts in many nations—from the Main Streets initiatives in Boston, which began in 1995,[1] and the Vital'Quartier policies in Paris, beginning in 2004,[2] to the program to improve high streets in Britain, which began in 2014.[3] These policies stake the life of the city on the vitality of local shopping streets, and offer financial incentives to individually and locally owned shops to share the risk. They are often carried out by public–private partnerships, including BIDs in Britain and a nonprofit foundation in Boston. But unlike BIDs in the U.S. and BIAs in Canada, under the British plan BIDs offer substantial, direct financial and technical aid to specific shops.

It is pointless to expect state policy to try to eliminate chain and discount stores and their low prices. It is, however, reasonable to ask large-scale retail chains to pay the real costs of their way of doing business. Because so many chain stores are built with large parking lots that encourage individual automobile transportation to and from the store, why not ask them to bear a larger part of the environmental costs by paying an extra local tax? Why not limit the size of the stores they want to open in cities, as New York City has done by changing the zoning laws, though in only one area of the city, the Upper West Side of Manhattan?[4]

Rent

From New York to Shanghai, rapid, massive rent increases are a perennial complaint of local shopkeepers on almost all of the streets we studied. Though no surveys conclusively document the impact of rent increases, or demanded increases, on the fortunes of local shops, we heard many anecdotal stories about store owners who had not been able to meet the financial demands of their landlords, and were forced to close—some, after many years of doing business on the street. Whose interests, exactly, does that serve?

Building owners certainly must make an adequate profit from renting commercial space, enough to pay their taxes and keep reinvesting in building maintenance. But as property owners, their service to the local economy—and local community—does not outweigh that of their commercial tenants. Local merchants often sustain neighborhoods in economically hard times, not because they are "better" people than chain store owners or even necessarily more committed to the neighborhood, but simply because they have nowhere else to go and no easy way of moving their capital elsewhere. As we see in Toronto, they may even sustain local residents, especially young people, by extending credit for purchases, hiring them, and teaching them how to run a small business. However, local shopkeepers are often the first to suffer when commercial gentrification allows building owners to increase rents by large amounts at the end of their lease.

Commercial space in New York is especially prone to this kind of cataclysmic rent increase. Though all global cities attract overseas direct investment in real estate, which tends to drive up prices for property, New York seems to experience more extreme booms and busts—especially, in recent years, a prolonged boom that drives both commercial and residential rents to astronomical heights. Unlike residential markets, which still have large numbers of apartments under some form of rent regulation, commercial property markets have no mechanism to protect tenants from unreasonable rent increases. When a commercial lease ends, the landlord can charge whatever they think the market will bear, which could be double or even ten times the rent that the current tenant is paying. This situation has accelerated the closing of local shops and led to a large number of evictions of commercial tenants (Jonas 2014).

In response, a small group of city council members who represent low- and middle-income neighborhoods that depend on small shops, and where many small shopkeepers live, has introduced a Small Business Jobs Survival Act, which would require arbitration between landlords and commercial tenants if they cannot agree on a rent increase. Though advocates argue that this measure is needed to keep small, individually owned businesses afloat, protect jobs, and sustain the vibrancy of communities, the real estate industry insists that it violates the U.S. constitution's right of private property. The de Blasio administration came into office in 2014 pledging to help small business owners, and both simplified and reduced some burdensome regulations on them. But they may not be able or willing to oppose the real estate industry, which persuaded the city council to drop a similar bill five years earlier.

It is possible that the extreme highs and lows of New York's commercial real estate market force tenants and landlords alike to absorb both abnormally low and abnormally high rents. Perhaps if rent increases were moderated by custom or law—if a balance, in this sense, could be achieved—building owners would be less likely to demand a huge increase to make up for a loss, or a perceived loss, of rental income.

If New York rent rises are worrisome, so is the situation in Shanghai. In Tianzifang, for example, rents are accelerating faster than many businesses can bear. Given the area's initial orientation toward artistic and innovative businesses, building owners set relatively high rents to begin with. But because of Tianzifang's growing reputation and commercial success, they started to raise rents arbitrarily and without much regard to signed leases.

As in New York, very high rent increases are already forcing commercial tenants to leave, including the famous artists whose galleries and studios anchored Tianzifang's early development. Though these tenants look to public officials for support, there is little evidence that the local state is either able or willing to regulate the rental market, even by enforcing the landlords' legal obligations.

In Amsterdam, however, the rent situation seems more moderate. Commercial spaces generally have five-year leases with an automatic five-year extension. During the second five years, the annual rent increase is pegged to the annual increase in the cost of living. At the end of a lease, the landlord and store owner negotiate, often through the informal mediation of their respective real estate agents, who talk about what the market rent is in the locality. True, this does not always lead to a mutually satisfactory resolution. When store owners cannot pay the rent demanded, they must leave. But unlike in New York or Shanghai, where market forces hold sway, culture in the Netherlands is traditionally attuned to compromise. The impressive longevity of most retail businesses on Utrechtsestraat, admittedly an upscale shopping street, suggests an interesting model for export.

Perhaps this sort of "compromise culture" would not work in other cities, where the state is less likely to disrupt the market than in Amsterdam. But other mechanisms could be used to achieve the same result.

Commercial leases, with gradual rent increases tied to inflation or increasing profitability, could be encouraged. In exchange, however, landlords would have to forgo huge increases when a lease ends. Or, as New York's Small Business Jobs Act says, rent increases could be arbitrated. In another possible option, they could be regulated, like many residential rents, by annual negotiations between representatives of building owners and commercial tenants. More dramatically, a windfall profits tax could be levied on massive increases in commercial rents, and revenues from that tax could be used to subsidize lower rents for either tenants or landlords, or both.

Alternatively, BIDs could subsidize rents for startup businesses in otherwise vacant storefronts. The Fashion District BID in Manhattan has discussed setting up such an incubator, and the Lower East Side BID tried it, also for startup clothing designers, in 2002. The Parisian Vital'Quartier program has done it to support types of businesses that are under-represented in a neighborhood, or that could "diversify" local commerce.

But there's that word again. In Paris as in New York, shops that are given preference for subsidized space in incubators are "gentrifying" businesses run by young, creative professionals. Indeed, in New York, the major incubator efforts are now directed toward technology startups, not retail stores. Even under different types of programs, local shopping streets in New York and Paris begin to look the same as in Amsterdam.

A truly innovative option would be for city governments to create a local carbon tax, to be paid by developers of shopping centers with big parking lots on the edge of the city. The tax would reflect the actual social cost of such businesses in the form of environmental harm, such as carbon pollution from automobile traffic to and from the stores, and economic burdens on the local state, such as the need to build and maintain roads. A part of the revenue generated could be dedicated to subsidizing rents on more environmentally sustainable local shopping streets for longtime store owners.

Too Small to Fail

In the end, "the market" is not, and cannot be, the sole and final arbiter of local shopping streets' survival. These are social spaces, and must be supported by policy for the public good. They are also a form of cultural heritage that sustains, and is sustained by, generations of city dwellers. Again, for this reason, local shopping streets demand support.

As observers of society, we must note that local shopping streets confirm the social and cultural embeddedness of economic activity. Though each store owner acts individually to meet "market forces," they also act collectively to create a sense of place. Despite being profit-maximizing economic actors, they are important social actors, offering a priceless "home" to many different people who pass through, and by, their doors.

The economic historian Karl Polanyi (1944) pointed out years ago that neither markets nor the economy as a whole can work without the strong regulatory framework established by the state, or outside of cultural norms that limit their sometimes cruel effects. We feel the same about local shopping streets. In their own way, as much as big financial corporations, they need life support. But in their case, where everyday actions support everyday diversity, they are just too small to fail.

Notes

1   See www.bmsfoundation.org (accessed January 16, 2015).
2   See www.semaest.fr/nos-realisations/vital-quartier (accessed January 16, 2015).
3   See www.gov.uk/government/policies/improving-high-streets-and-town-centres (accessed January 16, 2015).
4   See www.nyc.gov/html/dcp/html/uws/index.shtml, June 28, 2012 (accessed January 16, 2015).

References

Anderson, Kay. 1995. *Vancouver's Chinatown: Racial Discourse in Canada, 1875–1980*. Montreal: McGill-Queen's University Press.

Aytar, Volkan and Jan Rath, eds. 2012. *Selling Ethnic Neighborhoods: The Rise of Neighborhoods as Places of Leisure and Consumption*. New York: Routledge.

Duyvendak, Jan Willem. 2011. *The Politics of Home: Belonging and Nostalgia in Western Europe and the United States*. Basingstoke: Palgrave Macmillan.

Hall, Suzanne. 2012. *City, Street, and Citizen: The Measure of the Ordinary*. London: Routledge.

Jacobs, Jane. 1961. *The Death and Life of Great American Cities*. New York: Random House.

Jonas, Jillian. 2014. "City's Small Business Crisis Continues." July 24. Available at www.gothamgazette.com/index.php/government/5155-new-york-city-small-business-crisis-continues (accessed January 16, 2015).

Kallick, David Dyssegaard. 2015. *Bringing Vitality to Main Street: How Immigrant Small Businesses Help Local Economies Grow*. New York: Fiscal Policy Institute and Americas Society/Council of the Americas.

Kosta, Ervin. 2014. "The Immigrant Enclave as Theme Park: Culture, Capital, and Urban Change in New York's Little Italies." In *Making Italian America: Consumer Culture and the Production of Ethnic Identities*, edited by Simone Cinotto, pp. 225–43. New York: Fordham University Press.

Polanyi, Karl. 1944. *The Great Transformation*. New York: Farrar & Rinehart.

Robinson, Jennifer. 2011. "The Spaces of Circulating Knowledge: City Strategies and Global Urban Governmentality." In *Mobile Urbanism: Cities and Policymaking in the Global Age*, edited by Eugene McCann and Kevin Ward, pp. 15–40. Minneapolis MN: University of Minnesota Press.

Vertovec, Stephen. 2012. "'Diversity' and the Social Imaginary." *European Journal of Sociology* 53(3): 287–312.

Wessendorf, Susanne. 2014. *Commonplace Diversity: Social Relations in a Super-Diverse Context*. Basingstoke: Palgrave Macmillan.

Zukin, Sharon. 2010. *Naked City: The Death and Life of Authentic Urban Places*. New York: Oxford University Press.

Research Note

How to Put a Transnational Project Together

SHARON ZUKIN, PHILIP KASINITZ,
AND XIANGMING CHEN

A local shopping street seems to be the simplest of urban spaces, but it is a remarkably complex social production. And it changes all the time.

It may wear its history lightly, but the buildings and storefronts incorporate the stories of many men and women, extending back in time and around the world. Today the shopkeepers come from one distant region or another, yesterday they came from other places; years ago, they were children of farmers who came to the city to look for work. The stores on the street have also had many lives. Today they may be upscale shops, but last year, or maybe thirty years ago, they were bargain stores.

How can you discover all the lives of a shopping street? Are the changes we are so aware of today, changes produced by globalization and gentrification, universal?

To explore the mutual shaping of local streets and global trends, we created a transnational research project with six local teams of research partners. Two of us began by forming the New York team. Though both of us are urban sociologists and have written about New York, one has focused on gentrification, and the other, on immigration. We recruited another "local" partner, also an urban sociologist, who has lived in, and written about, Shanghai. The three of us worked closely together to develop the research project and edit each chapter of the book, and each of us has co-authored individual chapters.

We knew it was essential to recruit a transnational research team with the deep local knowledge that studying cities requires. Urban researchers need to understand local laws and policies, from zoning laws that dictate the uses of buildings in different districts of the city to plans for redevelopment that have been debated and then either enacted or discarded. They must know the local language to carry out interviews and read the media. It's good if they have a long memory to recall what specific streets were like, and what people said about them, in earlier times. Researchers' memories don't compensate for documentary sources, but their subjective knowledge helps them to understand the issues and social context that reshape a street's reputation over time, so that one street becomes very different from another.

We recruited Katherine Rankin, a geographer who had already done research on local shopping streets in Toronto, Jan Rath, a sociologist who writes about immigration to Amsterdam, Yu Hai, a sociology professor in Shanghai who often comments on urban development there, Takashi Machimura, an urban sociologist who has done extensive field work in Tokyo, and Keiro Hattori, an expert on environmental planning, also in Tokyo, who had done research on local shopping streets there. We were soon joined by Talja Blokland and Christine Hentschel, urban sociologists whom we knew in Berlin, and a number of graduate students in all six cities. Most important for the New York research team, we relied on a dozen students in the PhD program in sociology at the CUNY Graduate Center, with whom we did field work, planned conference presentations, and held weekly workshops.

Clearly, the cities and regions that we covered are not the only places where globalization and, to some degree, gentrification are important. But North America, Europe, and East Asia are now on a similar level of economic development. The cities have long histories as commercial centers. Most of them, especially in North America and Europe, have large percentages of transnational migrants in their population. Unlike in the Global South, local shopping streets in these regions tend to depend less on informal markets and more on formal regulation by the city government or local state.

Limiting our research project to six cities also responds to practical needs. Communications and procedures are quicker to coordinate in a small group. Making comparisons is easier with a small sample. Using two cities in each of three regions helps us, and helps readers, to see contrasts and similarities. But we have dared to hope from the outset that this work would only be the first round of research, and that a "Local Shops 2" will expand our project to more regions of the world.

Thanks to small grants from the Fund for the Advancement of the Discipline of the American Sociological Association and from the Russell Sage Foundation, we organized a workshop at the Graduate Center of the City University of New York that brought most of the research partners together, face to face, for the first time. With support from Fudan University, Trinity College, and the University of Amsterdam, we were able to meet in two more annual workshops in Shanghai and Amsterdam.

Selecting Issues and Sites

We began the first workshop with short PowerPoint presentations on a few likely research sites, chosen by the research team in each city. Viewing the photos and maps, and listening to the observations of the local research partners, led to intense discussions of what our common questions could be. Not surprisingly, globalization emerged as the primary issue. The second was the role of the local state in redevelopment and its response to national and transnational migration. Several streets also showed intriguing signs of gentrification, marked by the apparently universal ABCs: art galleries, boutiques, and cafés. Though these businesses reflected market-based gentrification on some streets, they had been prodded into place on other streets by the local state.

These three issues—globalization; local state action, especially in response to migration; and gentrification—formed our common research agenda. During the next four years, we investigated them on two local shopping streets in each of the six cities, selecting these streets for convenience, interest, and contrast.

In New York, we chose streets in two neighborhoods that had been considered ghettos for most of the twentieth century. In Shanghai, our streets were newly developed as shopping districts—one as a cultural zone and the other as an everyday shopping street—during the past thirty years. In Amsterdam, one street is upscale while the other is downscale. In Berlin and Toronto, both streets are in working-class to low-income neighborhoods with strong majorities of transnational migrants as both residents and shopkeepers. But in each of those two cities, one street is being gentrified. In Tokyo, the streets are quite different. Both are in middle-class residential areas, but one is a traditional, upscale shopping street and the other specializes in music clubs and curry restaurants with a hipster vibe.

The three annual workshops brought us together to visit half of all the research sites. Seeing other local shopping streets enabled us to see our own streets more clearly, to frame issues more broadly, and to look for the global agents and processes that structure local change. One of these processes is gentrification, which has become, as the geographer Neil Smith (2002) has written, a global strategy of inner city redevelopment. But in some cities, gentrification is associated with "diversity" and praised for fostering social integration. In other places it is condemned for displacing vulnerable migrants and exerting a homogenizing force.

Research Questions

This quick overview of our research issues introduces four big empirical questions.

- **Change:** How do local shopping streets change over time? When does one kind of business morph into another, why does one group of shopkeepers or shoppers replace another, how does the cultural ecosystem of the local shopping street adapt to big changes in the economy, society, and consumer culture?
- **Community:** How are social and cultural communities formed on the street? Do groups experience conflict or cooperation, or are they merely co-present with little interaction? Are local shopping streets spaces of "everyday diversity" where tolerance is learned and dignity is respected?
- **Government:** Are markets the only force that matters on local shopping streets? Or are these streets shaped by the invisible hand not only of the market but of the local state?
- **Homogenization:** Facing powerful forces of economic and cultural globalization, do all local shopping streets begin to look alike?

Research Methods

The research project faced two great challenges. First, where could we find empirical materials that would speak to these questions? Second, how could we relate thick description of each local shopping street to abstract social processes of globalization and gentrification? We started our time frame in 1980, when the current era of globalization began, and we continued to follow the twelve streets until 2014.

Residential District

To understand each local shopping street, its relative place in the city's social geography, and how it has changed since 1980, we began with the surrounding residential district. For each street, we collected data about the social class, income levels, and ethnic and national backgrounds of the local residential population, and how these factors have changed over time, using data points in 1980, 1990, 2000, and 2010. In the U.S., this data comes from the decennial national census. In other countries, data on population by area of residence is collected by different local and national state agencies.

Changes in Local Businesses over Time

It is hard to get information about changes in the retail landscape. In New York, no government agency collects data on which businesses occupy the same storefront over time. So the New York team consulted Cole's reverse telephone directories that are compiled by a private company and are available in the Science, Industry and Business Library of the New York Public Library. Rather than listing every person or business that has a land line alphabetically by family name, these directories list them by street address. Although reverse phone books have been an indispensable resource so far, they may become less useful in the future if more small business owners give up land lines for cell phones. Mobile phone numbers are not linked to specific addresses. If they were, these might be home rather than business addresses.

In the other five cities, we gathered information about changes in stores over time from either business directories or local government agencies which, unlike in New York, collect and store the names of businesses by address. This task sounds easy, but it takes persistence to track down the data and get permission to access it.

Whatever the data base, lists of stores may be limited or incomplete. Often it is impossible to guess the kind of products being sold from the name of the business. It is even harder to guess the affluence of the target market, or a shop's "ethnic character." Therefore, the largest category of businesses that we often found was "unknown." To deal with this data gap, we searched for each "unknown" business on the World Wide Web to try to find some indication of their products or reputation. Unfortunately, most small businesses leave no trace online or in social media.

Nevertheless, we did find important historical changes in retail landscapes. On the one hand, most local shopping streets show a remarkable continuity in the sectoral distribution of kinds of stores that remain there over time, from grocery stores, clothing stores, and dry cleaners, to barber shops and hardware stores. On the other hand, some kinds of businesses disappear (record stores, egg stores) while others rise (cell phone stores, nail salons). These changes in the business composition of a local shopping street tell a bigger story of structural change in the economy, which in turn shapes the narrative of how cities change.

The longevity of stores, that is, the number of years individual businesses survive, indicates something about the ecosystem of local shopping streets. Individually owned businesses that survive for many years at the same address show economic stability, which in turn supports the social and cultural reproduction of community. Shops that turn over quickly suggest a different story, one of a community's economic, and perhaps also social, marginalization.

Social Class, Ethnic Identity, and Hipsters

To update historical data about the business composition of each shopping street, we also carried out a "walking census." We walked up and down every block, noting the kinds of products sold in the storefront at each address, their general price level, and the shop's overall aesthetic. In the New York research team, we had many heated discussions about aesthetics, for characterizing them is endlessly fascinating, perplexing, and unavoidably subjective. A store's color scheme (primary colors or white walls) and lighting (fluorescent lights or vintage Edison bulbs), a restaurant's sound track (1940s jazz or Latin funk), and the quantity and arrangement of its stock (small and "curated" or big and messy): all of these are cues of "distinction" in the hierarchical sense described by the sociologist Pierre Bourdieu (1984). But it isn't easy to agree on how to describe and classify these cues. And not all of the research teams made it a priority to get the same data.

Efforts to categorize businesses by their "ethnic character" brought up similar issues of data that was either incomplete or misleading, compounded by our unwillingness to accept stereotypes. We didn't know how to describe the ethnic identity of stores that had vanished into the past. But today, signs written in different languages, products identified with different cultures and regions, and flags of different countries are strong indicators of ethnic and national identity, at least on the American and European shopping streets.

Besides ethnicity, two other cultural categories—hipsters and religion—were problematic. When you say "gentrification by hipster" everyone laughs, but everyone also knows what kind of a shopping street that phrase suggests. Though the specific components of "hipsters" are always changing, the terms "hip" and "trendy" are associated with a global toolkit of specific products, tastes, and brands. Businesses catering to these tastes, which historically have been identified with bohemian subcultures, are often concentrated geographically in specific streets and neighborhoods.

Similarly, most stores do not promote themselves on the basis of religion. But for a century following the great wave of immigration from Russia and Eastern Europe in the 1880s, one of our New York streets supported a spatial concentration of Jewish store owners, who served the daily needs of co-ethnics and co-religionists in the surrounding neighborhood, the Lower East Side. Today, our other New York street, in Central Brooklyn, hosts a halal butcher shop and about a dozen halal restaurants, a spatial concentration of Muslim-owned stores that serves a newer wave of transnational migrants from West Africa and South and Central Asia.

Interviews with Store Owners

We explored ethnic, religious, and lifestyle identities, and the communities that form among businesses that cater to them, in semi-structured interviews with business owners, shoppers, and local officials of government agencies and representatives of public–private associations, especially the organization that represents business and building owners. In Amsterdam, this is the street (or merchants') association. In New York, it is the business improvement district (BID), and in Toronto, the business improvement area (BIA). We also spoke with police officers and local residents. But we relied mainly on interviews with business owners and managers because we considered them to have the most direct information about the street. Unlike shoppers and building owners, their work requires them to spend all day on the street.

In interviews with a representative sample of business owners on each of our research sites, we asked about the history of each business and the owner's life story. We asked how the business owners discovered the street and decided to locate their business there, what their experiences had been, and how the street had changed since they arrived. We wanted to know whether they own the building or rent the store. We also asked who their customers are, and how the owner tries to attract them. We inquired about the owner's ethnic background and national origins, education, and business history. And we asked about their interactions with other business owners, on the one hand, and with the local business association and local government, on the other.

Some interviews were more comprehensive than others. In New York, especially, we found that many small shopkeepers do not have time or patience to speak at length with a visitor who is clearly not going to buy anything, whether the stores deal in sneakers or works of art. By contrast, in Amsterdam, Tokyo, and Shanghai, business owners tended to speak freely and at length.

Ethnographic Observations

It's easiest to do ethnography if you live or work on your research site. In the six cities in our study, the researchers only lived near three of the twelve research sites. But we observed each street on different days of the week and at different times of day, and tried our best to spend time in different types of businesses where we could observe the customers and interaction between them and the owners.

We looked at how people walk through each street, how many are there at different times of day, and what they do. We sat in Dunkin' Donuts, music bars, and hipster cafés. We bought groceries and takeout sandwiches. We used the services of dry cleaners and opticians. We sat and chatted with the barber even if we didn't get a haircut.

On Fulton Street in Brooklyn, our student researchers got suspicious looks from passers-by, and several times people asked them why they were walking up and down the street making notes. (This was when the students were carrying out our walking census of stores.) We think this reaction responds to an ongoing process of gentrification, and racial and ethnic change, of which local residents are well aware and to some degree distrustful. It emphasizes how important it is to gain the trust of potential research subjects, which requires an ethnographic commitment to spending time on the street in order to be seen, to some degree, as a "local." Sometimes our student researchers visibly shared the racial or national background of their interview subjects, and sometimes that mattered.

Internet Presence

We completed our research by searching for each street, and many businesses, on the Internet, to see how they are represented, by whom, and where. One of our early findings was that new, trendy shops are heavily promoted on social media, while most traditional, inexpensive, immigrant owned shops are not. Perhaps this digital divide is not surprising; however, it highlights a difference between globally constructed, locally bounded cultural communities that depend on word of mouth recommendations and ethnic and religious ties, and other globally constructed, but more locally mobile cultural communities that connect through social media and common consumer tastes.

Writing Process

The writing process extends the research methodology known as grounded theory. Truths emerge in the writing, yet we choose concepts to try to capture the social realities of our subjects' lives from the vocabularies we have already learned. While our writing uses terms like social class, ethnic identity, and gentrification to build narratives about localities, we also use citizenship, diversity, and belonging to create narratives of globalization.

Some of these terms reflect the interests of our research partners who are primarily interested in issues of social diversity and immigration. Others relate to interests in the changing texture of urban experience and structural economic and political transformation. As this suggests, differences in our research priorities made for variations in the way each research team approached the project and wrote up their results.

Meeting in the three annual workshops helped us to stay on the same general track. The editorial base in New York suggested a uniform format for all the chapters, including the "shopkeepers' stories" that show the intersection of global processes, individual biographies, and the collective history of the street. These stories, in fact, confirm the benefits of collaboration, for the idea of creating them came from the students in our New York research team.

The three book editors co-authored three of the "city" chapters and edited the other three. We also wrote the introduction and conclusion. We think it is clear that, just as a local shopping street has many "authors," so have all of our collective efforts led to a multi-sited, multi-authored book.

Acknowledgments

In each chapter, the local research team thanks all the interviewees, supporting universities and organizations, and outside experts who have generously given their help and time. In our case, we begin by thanking all of the research partners. Without their willingness to share our vision, this transnational project would never have got off the ground.

We thank Roberta Spalter-Roth, former director of the Fund for the Advancement of the Discipline of the American Sociological Association, and Eric Wanner, former president of the Russell Sage Foundation, for enabling us to bring the research teams together for the first time. Their support for this project, which was located off the radar screen of their organizations' usual concerns, was crucial to pursuing a new set of issues in an innovative way.

Because this project was not funded by any large grants, we are especially grateful for the support of the University of Amsterdam, Fudan University, and Trinity College, which enabled us to sustain the transnational collaboration. These institutions made the "global community of scholars" a reality.

We deeply appreciate the encouragement and support of Steve Rutter, former social sciences publisher at Routledge in New York, and Margaret Moore, senior editorial assistant, who helped us to bring the book manuscript and its many visual images to life. They share our vision to use these tools to create a multi-media, multi-dimensional community of readers, who care about reconciling the local and the global in their city's social life.

Finally, we thank, again, the students in our research team in the PhD Program in Sociology at the Graduate Center of the City University of New York. You allowed us to teach you how to do research, and we learned much from the experience. And we are grateful to undergraduate research assistants at the Center for Urban and Global Studies at Trinity College for creating and maintaining a companion website for the book.

References

Bourdieu, Pierre. 1984. *Distinction: A Social Critique of the Judgement of Taste*, trans. Richard Nice. Cambridge MA: Harvard University Press.

Smith, Neil. 2002. "New Globalism, New Urbanism: Gentrification as Global Urban Strategy," *Antipode* 34, 3: 427-50.

Brief Biographies of Research Partners

Talja Blokland

Talja Blokland is professor of urban sociology at Humboldt University in Berlin. Her most recent books include, with A. Harding, *Urban Theory* and, with Mike Savage as co-editor, *Networked Urbanism*. She specializes in the study of ghettos and urban poverty, the city as a site of resources and relational understandings of urban inequalities.

Xiangming Chen

Xiangming Chen is the dean and director of the Center for Urban and Global Studies and Paul E. Raether distinguished professor of global urban studies and sociology at Trinity College in Hartford, Connecticut. He is also a distinguished guest professor in the School of Social Development and Public Policy at Fudan University in Shanghai. His (co-)authored and co-edited books include *As Borders Bend: Transnational Spaces on the Pacific Rim*, *Shanghai Rising: State Power and Local Transformations in a Global Megacity*, and *Rethinking Global Urbanism: Comparative Insights from Secondary Cities*.

Iris Hagemans

Iris Hagemans has an MA in urban studies from, and has been a junior lecturer at, the University of Amsterdam. She is the co-author of "'Gentrification without Displacement' and the Consequent Loss of Place," about residents of secure low-income housing in Melbourne, Australia, and is now working as a GIS specialist.

Keiro Hattori

Keiro Hattori is a professor in the Faculty of Economics at Meijigakuin University in Tokyo. Active in the fields of urban planning and sustainable development, he has published books and reports including *Curitiba, Human Oriented City, The Great Sin of Road Policy*, and *Town Planning for Teenagers*, and translates books from English into Japanese. Also a licensed city and regional planner, he served as a site planner and urban designer on the Cyberjapa Planning Project in Malaysia in 1996–97.

Anke Hendriks

Anke Hendriks has an MA in urban sociology from the University of Amsterdam and is a junior lecturer there.

Christine Hentschel

Christine Hentschel is visiting professor of international criminology at the University of Hamburg. Between 2011 and 2014, she was a postdoctoral fellow in the department of urban sociology at Humboldt University. The author of *Security in the Bubble: Navigating Crime in Urban South Africa*, she also received the Heinz-Maier Leibnitz Award of the German Research Foundation for her work on postcolonizing Berlin.

Kuni Kamizaki

Kuni Kamizaki holds a Master's degree in Planning with a Certificate in Community Development from the University of Toronto. He was a research assistant with the Commercial Gentrification project at the Department of Geography and Program in Planning at the University of Toronto from 2009 to 2013. Currently he is co-investigator (with Katharine Rankin) for research on neighborhood-based local economic development activities in Toronto. He is also Research and Community Economic Development Coordinator at Parkdale Activity Recreation Centre (PARC), a multi-service community development organization.

Philip Kasinitz

Philip Kasinitz is presidential professor of sociology at the Graduate Center of the City University of New York. His books include *Caribbean New York: Black Immigrants and the Politics of Race, Metropolis: Center and Symbol of Our Time, Becoming New Yorkers: Ethnographies of The New Second Generations,* and *The Urban Ethnography Reader.* He is co-author of *Inheriting the City: The Children of Immigrants Come of Age*, which received the 2010 Distinguished Publication Award from the American Sociological Association.

Sunmee Kim

Sunmee Kim is a PhD candidate in the Graduate School of Social Sciences, Hitotsubashi University, Tokyo. She has published "The Role of Art in Revitalizing Tokyo's Inner City: Between Gentrification and Diversification of Local Culture," and is the co-author of *Anti-Nuclear Social Movements after the Accident at Fukushima*, both in Japanese.

Takashi Machimura

Takashi Machimura is professor of sociology at Hitotsubashi University, Graduate School of Social Sciences, in Tokyo. His recent books include *Structure and Mentality in Developmentalism: Dam Construction in Postwar Japan*, and, with Sunmee Kim, *Anti-Nuclear Social Movements after the Accident at Fukushima* (co-author), both in Japanese. He has also edited *Exclusion and Resistance in Urban Space* and co-edited *What is Civic-Initiative Society?: Conflicting Public Sphere in the Decision Making Process of Aichi EXPO 2005*, both in Japanese, and published "Symbolic Use of Globalization in Urban Politics in Tokyo," in English.

Heather McLean

Heather McLean is a post-doctoral researcher in the University of Glasgow's Geography Department, working on the role of community-engaged arts in planning, development and redevelopment strategies. This work is influenced by her involvement in Toronto-based arts collectives as a performer and activist.

Katharine N. Rankin

Katharine N. Rankin is a professor in the Department of Geography and Program in Planning at the University of Toronto. She is the author of *Cultural Politics of Markets: Economic Liberalization and Social Change in Nepal*, as well as articles on the politics of planning and development, comparative market regulation, feminist and critical theory, neoliberal governance, and social polarization.

Jan Rath

Jan Rath is professor of urban sociology, and associated with the Institute for

Migration and Ethnic Studies (IMES) and the Center for Urban Studies, at the University of Amsterdam. His co-authored and co-edited books include *Immigrant Businesses: The Economic, Political and Social Environment*, *Unravelling the Rag Trade: Immigrant Entrepreneurship in Seven World Cities*, *Immigrant Entrepreneurs: Venturing Abroad in the Age of Globalization*, *Tourism, Ethnic Diversity, and the City*; *Ethnic Amsterdam*; *Selling Ethnic Neighborhoods*, and *Immigration and the New Urban Landscape: New York and Amsterdam*.

Hai Yu

Hai Yu is professor of sociology in the School of Social Development and Public Policy at Fudan University in Shanghai. His authored and edited books include *A History of Western Sociological Thought, A Study on Community Development in Shanghai, SARS: Globalization and China, On Shanghai Students' Spirit of Patriotism*, and *A Reader on Urban Theory*. His articles on Tianzifang have appeared in the *Nanjing Journal of Social Sciences* (in Chinese).

Xiaohua Zhong

Xiaohua Zhong is assistant professor of sociology at Tongji University and a consultant on urban heritage conservation and sustainable development in Shanghai. Her research focuses on urban community governance, heritage conservation, and creative industry clusters. She is currently involved in a collective archival project on Tianzifang which includes oral history.

Sharon Zukin

Sharon Zukin is professor of sociology at Brooklyn College and the Graduate Center of the City University of New York, and was a visiting professor in the Center for Urban Studies at the University of Amsterdam in 2010–11. She has written three books about New York City: *Loft Living, The Cultures of Cities*, and *Naked City: The Death and Life of Authentic Urban Places*, as well as *Point of Purchase: How Shopping Changed American Culture*. She won the C. Wright Mills Award from the Society for the Study of Social Problems for her book *Landscapes of Power: From Detroit to Disney World*.

Index

220